Not in Our Classrooms

NOT IN OUR CLASSROOMS

Why Intelligent Design Is Wrong for Our Schools

Edited by Eugenie C. Scott and Glenn Branch

Beacon Press, Boston

BEACON PRESS
25 Beacon Street
Boston, Massachusetts 02108-2892
www.beacon.org

Beacon Press books
are published under the auspices of
the Unitarian Universalist Association of Congregations.

Printed in the United States of America

09 08 8 7 6 5 4 3 2

This book is printed on acid-free paper that meets the uncoated paper
ANSI/NISO specifications for permanence as revised in 1992.

Composition by Wilsted & Taylor Publishing Services

Library of Congress Cataloging-in-Publication Data
Not in our classrooms : why intelligent design is wrong for our schools /
edited by Eugenie C. Scott and Glenn Branch. — 1st ed.
p. cm.
ISBN-13: 978-0-8070-3278-7 (pbk. : alk. paper)
ISBN-10: 0-8070-3278-6 (pbk. : alk. paper) 1. Intelligent design (Teleology)—
Study and teaching. 2. Creationism—Study and teaching. 3. Evolution (Biology)—
Study and teaching. 4. Evolution—Study and teaching. 5. Religion and science.
I. Scott, Eugenie Carol. II. Branch, Glenn.

BL263.N68 2006
231.7'652—dc22 2006016409

Credits
"Evolution in the Classroom" by Brian Alters is adapted from Defending Evolution in the Classroom: A Guide to the Creation/Evolution Controversy by Brian Alters and Sandra Alters (Boston: Jones and Bartlett, 2001).

Portions of "Defending the Teaching of Evolution" appeared previously in Glenn Branch, "The Battle over Evolution: How Geoscientists Can Help," The Sedimentary Record 3, no. 3 (2005): 4–8.

Contents

Foreword

REV. BARRY W. LYNN

Just weeks after the legal team from Americans United for Separation of Church and State, the Pennsylvania affiliate of the ACLU, and the law firm of Pepper Hamilton LLP won a stunning victory in the Dover, Pennsylvania, intelligent design case, another creationism conflict reared its ugly head. Right after January 1, 2006, we at Americans United started to get phone calls from citizens of the small mountain town of Lebec, California.

The school board for the El Tejon Unified School District had approved a between-semesters intersession class called Philosophy of Design. Hadn't anyone read the ubiquitous press accounts about Dover? They had, but they also read one brief passage in the 139-page opinion in which the judge opined that "ID should continue to be studied, debated, and discussed... our conclusion today is that it is unconstitutional to teach ID as an alternative to evolution in a public school science classroom." A dim light bulb went off in the minds of three of the five board members: if you call a class *philosophy,* it is not in the science department, and therefore the whole conflict of the Dover case could be avoided.

I didn't think so. This class was being taught by a woman who made it mighty clear in her syllabus and course description that she would be offering a biblical attack on the evidence for evolution and the age of the Earth. It was a kind of infomercial for creationism with academic credit attached. The teacher later told the press "this is the class that the Lord wanted me to teach."

We needed to act quickly. Our lawyers filed a motion in federal court on behalf of eleven parents in the district for a temporary restraining order to stop the class, which began on January 3, from continuing. This was filed on a Wednesday, but the board was meeting just two days later, which was four days before the hearing on our case would begin. At this point, something exciting and eye-opening happened: The community itself got interested in the issue. The local paper devoted a full and unparalleled six pages of letters to the editor on the now suddenly controversial course. We called some of our members in the community (we had a mere seven contacts in the area), and those members—all of whom have friends and contacts in other organizations in the area—showed up in force and in earnest at that Friday meeting, insisting that the board change direction before the lawsuit began. The board was convinced, and a settlement was reached over the weekend: an early termination of the course and a court-enforceable promise never to offer a similar course again. It was the right thing to do and it potentially saved the community millions of dollars in legal costs (these cases are that expensive; even more so if the school loses).

No matter how credible the scientific evidence is in the rest of this book; no matter how clear the constitutional arguments; no matter how well crafted the explanations that evolution and religious faith are not in conflict—this is not a battle that will go away soon. Dover, Pennsylvania, and Lebec, California, are unlikely to be the last battlefields (and you are not reading this volume for its historical interest). During 2005, antievolution legislation was introduced in at least a dozen states. In recent months, the governors of Texas and Kentucky have expressed their belief that intelligent design has a place in their state schools. State boards of education in Kansas, South Carolina, and Ohio have been grappling with possible curriculum changes on this topic as well.

Antievolutionism will continue to be a popular political cause for the champions of the so-called Religious Right in their ongoing crusade to convince the nation that there is a war against people of faith. When you look at the cover of *Persecution*, the wildly popular book by David Limbaugh, you see a lion, harking back to the days of

Roman emperors who cast Christians into arenas with these kings of the jungle. Some people believe there is indeed a comparable battle today, and it is being fought in America's public schools. In their view, the schools are filled with anti-Christian materials (which include biology books with chapters on evolution) and the censorship of all other ideas (including, of course, intelligent design).

Some of the rhetoric is very clever: most Americans don't approve of censorship; they don't want to support the persecution of people of faith. Well, neither do the authors of this book. We do, however, need to be able to explain why censorship and persecution are not in fact occurring in the science classrooms of America.

The proponents of the latest tactical assault on evolution simply invent a new spin to describe their position or find new legal attacks. The rhetoric is designed to cover up the unquestionably religious motivation they have. But the legal maneuvering to date has never actually prevailed. We can expect, though, that legal challenges, just like the legislative and administrative initiatives mentioned above, will continue. Already on the horizon are efforts to refine previously unsuccessful challenges that claim teachers are being denied their academic freedom if they are not allowed to interject creationism into their discussions of evolution. Several courts have discounted earlier versions of this approach by noting that public school teachers are not hired to create curriculum in conflict with state policies nor are they authorized to have unfettered control over what is taught in their classrooms.

I've been involved in various activist causes since I was in high school myself (which only creationists believe was when dinosaurs still roamed the Earth). I get up in the morning intending to win whatever battle I'm a part of, not necessarily that day, but sometime soon. Those of us struggling to preserve sound science education in America have some real challenges, but the issue is too vital to give in to despair or defeatism. "We the people" have always made a difference, and in what Dr. Martin Luther King Jr. called "the arc of history," we are bending it toward justice. I think we are also bending it against ignorance. Indeed sometimes the very best teaching moment is in the midst of local fights over an issue. The story of Lebec illustrates particularly well some of the ways in which community residents can work together to defeat intelligent design pedagogical experiments. This is the time that the media and community groups start to pay attention.

The media wants sound bites for the evening news or the morning paper, but they also are genuinely interested in presenting the issue in depth. We have to be able to help them in both ways.

Just as intelligent design supporters use inflammatory words like *censorship* and *persecution* early and often, we need to have responses that are quick and to the point: how we want top-quality science education and know that it is illegal to teach one religious viewpoint in a school. We need to avoid making references to the specific religious viewpoints of supporters unless they have already done so. In the Lebec case, the teacher was the wife of a local Assemblies of God minister whose denomination opposes evolution. We cautioned against assuming that she thought precisely like her husband, until, of course, she gave public statements confirming that she did.

This is also an opportunity to think creatively about the range of people and organizations that may be helpful. Intelligent design represents a religious viewpoint, but this does not mean that all leaders of the faith community support it; in fact, many do not see any conflict between the evidence for evolution and their own theological views about divine purpose in the universe. This was even the strong view of the late Pope John Paul II. In Utah's recent legislative debate on this topic, many of the legislators in both political parties who were members of the Church of Jesus Christ of Latter-Day Saints opposed efforts to water down discussion of evolution. This is a healthy debate going on in many denominations and faith groups, and locating the clergy on your side early is particularly helpful.

In addition, business leaders are also concerned about the quality of education and some will find it in their best interest as future employers to be sure that sound science education is a part of the school curriculum. Dogma doesn't build better medical devices; good science does.

I have had serious conversations with some science educators who are not convinced that they should get involved in community debates. Many of them properly see the ID proponents as promoters of a pseudoscience that is barely worthy of discussion. Unfortunately, those advocates sometimes have scientific credentials and use phrases like *irreducible complexity* that certainly sound scientific. If no one rebuts these ideas in plain English, listeners may presume that the ID claims go unanswered because they are true. Don't be embarrassed to encourage scientists (as well as any lawyers who are involved) to go through dry runs of public appearances to make sure they are speaking in ways that lay audiences can understand.

The book you have in your hands is an excellent resource to deal with the attack on evolution, which is a surrogate, and indeed a wedge, for a wide-ranging crusade against the scientific integrity of the public education system in America. Understanding that, as well as understanding what the claims of intelligent design proponents represent, will, I hope, help all of us together win this extraordinarily important battle.

1: The Once and Future Intelligent Design

EUGENIE C. SCOTT

In the Beginning

In the United States, evolution became widely accepted by the scientific community around the turn of the twentieth century. It thereafter began to be included in college and secondary school textbooks. The late nineteenth century was not a period of extensive religious hostility to evolution, partly because of the promotion of evolution by American scientists who were active church members. It was not until the twentieth century that the U.S. antievolution movement became organized, active, and effective. Three trends converged to produce the first major manifestation of antievolutionism in the twentieth century: the growth of secondary education, the appearance of Protestant fundamentalism, and the association of evolution with the ideas of social Darwinism and eugenics, which became unpopular after World War I.

MORE SECONDARY SCHOOL STUDENTS

Although textbooks at the turn of the century included evolution, few students were exposed to the ideas contained in these books: in the late nineteenth century, high school education was largely limited to urban dwellers and the elite. In 1890, for example, only 3.8 percent of children aged 14 to 17 attended school—about 202,960 students.[1] But high school enrollment approximately doubled during each subsequent decade, so that by 1920, there were almost 2 million students attending high school. The practical effect of this was that more students were being exposed to evolution—and parents who felt uneasy about the topic for religious or political reasons rallied around the politician William Jennings Bryan to protest the teaching of evolution to their children.

FUNDAMENTALISM

The fundamentalist movement in American Protestantism is named for a theological perspective developed during the first few decades of the twentieth century. It was encapsulated in a series of small booklets collectively called *The Fundamentals*, published between 1910 and 1915. Fundamentalists believed the Bible to be inerrant—that is, without error—and its passages were to be taken literally, not "interpreted."

Financed by millionaires who had founded the Bible Institute of Los Angeles, now Biola University, millions of copies of *The Fundamentals* booklets were printed and distributed "free of charge, to every pastor, professor, and theology student in America."[2] Some of the authors of the booklets rejected evolution, but some accepted various forms of theistic evolution, the view that evolution is the method used by God to bring about the current diversity of living things. However, the fundamentalist position hardened into antievolutionism fairly quickly. Fundamentalists became the ground troops for the campaign to rid schools of evolution. They were motivated by religious sentiments and also by a concern that evolution was the source of many negative and even corrosive social trends.

EVOLUTION AS A SOCIAL EVIL

The second decade of the twentieth century was a time of considerable social and psychological unrest. The appalling death, brutality, destruction, and devastation of World War I led many to conclude that civilization itself had failed. Conservative Christians sought a solution in a return to biblical authority. Their views were reinforced by the fact that Germany had been the main source of both theological Higher Criticism (viewed as an attack on religion) and militarism (viewed as an attack on civilization).[3]

Conservative American Christians believed German militarism, theories of racial superiority, and eugenics arose from the acceptance of evolution by Germany at the end of the nineteenth century. In reality, German views of evolution were quite different from those of Darwin, largely rejecting natural selection as a mechanism of change, biological or societal. In the early twentieth century, evolution was also credited with providing the foundation for laissez-faire capitalism—as robber barons of the late nineteenth and early twentieth centuries sometimes claimed that natural selection justified their exploitative labor policies and cutthroat business practices.

Thus fundamentalists, led by the famous progressive politician and champion of the working man William Jennings Bryan, had many reasons to oppose the teaching of evolution to their children, whether or not these reasons were justified. Beginning in the early 1920s, several state legislatures took up Bryan's call to outlaw the teaching of evolution, and finally, on March 23, 1925, the Tennessee legislature passed the Butler Act. This set in motion events that would culminate in the Trial of the Century.

The Scopes Trial and Its Aftereffects

The Butler Act prohibited the teaching of "any theory that denies the Story of Divine Creation of man as taught in the Bible, and to teach instead that man has descended from a lower order of animal." The American Civil Liberties Union advertised for a plaintiff to challenge the suit, and the citizens of Dayton, Tennessee, offered John Scopes, whose name is now forever tied to the American creationism/evolution controversy.

The Scopes trial is familiar Americana: the titanic battle between two legal giants, Clarence Darrow and William Jennings Bryan; the hucksterism of the Dayton businessmen; the vicious if entertaining accounts of the trial by H. L. Mencken; the image of a boyish John Scopes looking slightly bewildered before all the cameras and microphones. What people sometimes overlook, however, is that Scopes lost: the Butler Act stayed on the books and it remained illegal to teach evolution in Tennessee for about forty more years.

The legal strategy of banning evolution by statute fell out of favor, however. Although Mississippi and Arkansas passed antievolution laws similar to the Butler Act in 1926, antievolution laws were defeated in Oklahoma, Missouri, West Virginia, Delaware, Georgia, Alabama, North Carolina, Florida, Minnesota, and California.[4] But even without formal banishment, evolution disappeared from the secondary school curriculum because of economic pressure. In the South, teachers and parents who chose textbooks in local districts preferred ones that slighted evolution, so textbook publishers were quick to remove, downplay, or qualify evolution to make sales. Books tailored for the southern markets were of course also sold elsewhere, and evolution disappeared from textbooks all over the nation.[5] The curriculum is shaped by textbook coverage, and since evolution was absent from the textbooks, it quickly disappeared from the classroom.

By 1930, only five years after the Scopes trial, an estimated 70 percent of American classrooms omitted evolution,[6] and the amount diminished even further thereafter. From the 1930s until the late 1950s, there was no need for a creationist movement because evolution was not being taught. It was not until evolution returned to the high school classroom in midcentury that creationists were prodded into action, and the creationism/evolution controversy sprang back to life as creationists lashed back at the reintroduction of evolution in American schools.

The Return of Evolution

After having been ignored in the precollege curriculum for almost thirty years, evolution came back into high school textbooks in a big way. The late 1950s saw a boom in federal expenditure for science education. Model textbooks in physics, chemistry, geology, and biology were commissioned. The scientists and master teachers at the National Science Foundation–funded Biological Sciences Curriculum Study (BSCS) ignored tradition and composed textbooks that reflected science as it was taught at the university level, including evolution, ecology, and human reproduction. In 1963, the first three BSCS textbooks were released, and all of them included evolution as a prominent theme.[7]

Partly because these new textbooks carried the stamp of approval of the NSF, but also because they were so much more interesting and up-to-date than existing books, school boards and textbook selection committees were eager to adopt them. Once the BSCS books began selling, commercial publishers started producing books in the same mold.[8]

The halcyon days for creationists of a high school curriculum *sans* evolution were over. How to combat evolutionism? Since the 1700s, some supporters of a literal interpretation of the Bible had argued that scientific evidence existed to support their views, but such arguments had diminished considerably after Darwin's *Origin of Species*. Now, in the middle of the twentieth century, those views were revived, partly in response to the increasing presence of evolution in textbooks and in the curriculum. If students were going to be taught evolution, antievolutionists argued, students also should be exposed to a biblical view. Creationists reasoned that if creationism could be presented as an alternative *scientific* view—creation science—then it would deserve a place in the curriculum. No one was more important in shaping this approach than Henry M. Morris.

Creation Science and Henry M. Morris

The late Henry M. Morris is widely considered to be the father of the twentieth-century movement known as creation science. Morris, a former hydraulic engineer, began his career as a creationist in 1946 with the publication of his first book, *That You Might Believe*, while he was still in graduate school. The book and its successor, *The Bible and Modern Science*, proclaimed a recent six-day (with twenty-four hours per day) creation, and a literal, historical Flood. These views were of course based on literal interpretations of Genesis, but the additional claim was made that special creationism can be supported by the facts and theories of science. Although both of these books are still in print and continue to sell, the modern creation science movement crystallized in 1961 with the publication of Morris's book *The Genesis Flood,* written with the theologian John C. Whitcomb.[9]

Like Morris's previous works, *The Genesis Flood* argued that most modern geological features could be explained by Noah's Flood, a view that had originally been popularized by the early twentieth-century Seventh-Day Adventist geologist George McCready Price.[10] Termed Flood geology, this view became the core of the new movement called creation science. The book's mix of theology and science is characteristic of creation science, and it continues to be widely read in evangelical and fundamentalist circles. *The Genesis Flood* proposed scientific evidence that Earth was less than ten thousand years old, and that evolution was therefore impossible. Fundamentalists were eager to claim scientific support for their religious views and use it to "balance" the teaching of evolution.

Morris worked tirelessly to strengthen the evangelical anti-evolutionist movement. To promote scientific research supporting the young age of Earth and the universe, the special creation of all living things, and Noah's Flood, he helped to found the Creation Research Society (CRS) in 1963, soon after the publication of *The Genesis Flood*. The *Creation Research Society Quarterly* (CRSQ) began publishing shortly thereafter, in 1964. This provided the scientific veneer: creation science now had its own journal.

TENETS OF CREATION SCIENCE

Creation science reflects the Christian theological view of special creationism. It holds that God created all things in their present form. The universe, therefore, did not develop gradually over time. Regarding living things, special creationism claims that God created plants and animals as separate "kinds." If evolution's dominant metaphor is a tree of life branching through time, the image brought to mind with special creationism is of a lawn, with each blade of grass being a separately created kind. In special creationism, living things do not share common ancestors. Similarly, creation science proponents profess that the universe came into being in its present form, and that living things are separately created kinds exhibiting limited genetic variability. There can be some limited evolution within kinds (such as the cat kind radiating into lions, tigers, pumas, bobcats, house cats, and so on) but no common ancestry of kinds. Evolution, of course, sees all living things as ultimately connected through a genealogical relationship; common ancestry is the fundamental difference between special creationism and evolution.

Creation science argues that there are only two views, special creationism and evolution; thus, arguments against evolution are arguments in favor of creationism. Literature supporting creation science is based on alleged examples of evidence against evolution, which are considered not only proof against evolution but also positive evidence for creationism. Understandably, there is nothing in the creation science canon providing a positive scientific case for the sudden emergence of the universe in its present form at one time, let alone for its specific doctrines of a six-thousand-year-old Earth and universe, the occurrence of a worldwide flood responsible for the fossil record and geological features such as the Grand Canyon, and the impossibility of evolution except within sharp limits.

CREATION SCIENCE EXPANDS

In 1972, Henry Morris and others founded the Institute for Creation Research (ICR) as a division of the Bible-based Christian Heritage College. ICR became an independent institution in 1980, moving from the San Diego suburb of El Cajon to nearby Santee, California, where it currently is housed in two large buildings. Now grown to an institution with a staff of more than forty, the ICR began as and remains the flagship antievolution ministry.

The ICR has grown steadily since its inception, taking pride in always ending the year in the black and never borrowing money for its building projects. To promote creation science, ICR conducts extensive outreach to churches and individuals. The ICR maintains the Museum of Creation and Earth History at its Santee headquarters, which attracts an estimated twenty thousand people every year.[11] Remodeled in 1992, it currently reaches thousands of schoolchildren each year, most of them homeschooled or attending Christian schools. Because of the religious orientation of the museum, few local public school teachers take their students to the ICR museum. The museum presents a journey through the seven days of creation, mixing biblical with scientific references. True to Morris's concern with Flood geology, there is a Noah's Ark diorama, presenting calculations of how many animals could have been housed on the Ark.

Many other creation science museums are operating or are in the planning stages. The most ambitious creation science museum is scheduled to open outside of Florence, Kentucky. Built by the international creation science ministry Answers in Genesis (AIG), perhaps now the largest creation science organization in the United States, the museum plans extensive exhibits on the six days of creation, a planetarium, picnic areas for tourists, and other attractions.

In addition to the ICR and AIG, other national young-Earth creationist organizations have heeded the creation science message of Henry Morris. Creation science is still the dominant form of antievolutionism in the United States.

Equal Time for Creation and Evolution

By the early 1960s, the concept of evolution was returning to science textbooks and classrooms after having been largely absent since the 1930s. It is not coincidental that Whitcomb and Morris's *The Genesis Flood* was published in 1961 and the ICR was founded a few years later: the increased exposure of public school students to evolution was a cause for alarm among conservative Christians. The renewed textbook emphasis on evolution generated conflicts in states with antievolution laws; teachers who wished to teach modern science would be breaking the law if they did so.

In 1965, Arkansas, Tennessee, Louisiana, and Mississippi still had Scopes-era antievolution laws on their books.[12] But that year, the Arkansas Education Association (AEA) decided to challenge the state's antievolution law, partly because the presence of evolution in textbooks put teachers on a collision course with the law. Rather than seeking a Scopes-style teacher defendant who would be prosecuted for breaking the law, the AEA instead challenged the law itself with a teacher plaintiff, Susan Epperson, who sought to teach evolution legally.[13] The trial itself was very short, taking only about two hours; the judge ruled that the antievolution law was unconstitutional. But to the surprise of Epperson and the AEA, the Arkansas Supreme Court reversed the lower court's ruling in a two-sentence decision in 1967.

On appeal, the U.S. Supreme Court upheld the initial decision, ruling in 1968 in *Epperson v. Arkansas* that the antievolution law was unconstitutional because it "selects from the body of knowledge a particular segment which it proscribes for the sole reason that it is deemed to conflict with a particular religious doctrine." (The ruling is explained in further detail in chapter 4.) Finally, forty-three years after the Scopes trial, it was unconstitutional to ban the teaching of evolution.

The Arkansas and other antievolution laws had hardly ever been enforced, so the effect of the *Epperson* decision was largely psychological. But if evolution could not be banned, how could children be protected from it? Keeping evolution out of the classroom was obviously not possible, with evolution included in most textbooks. Teaching the Bible along with evolution was one solution, but it quickly ran afoul of the First Amendment of the Constitution, which sets forth freedoms of religion, speech, and assembly. The Religion Clause reads, "Congress shall make no law respecting the establishment of religion, nor inhibiting the free exercise thereof." The Establishment Clause prohibits the state from promoting religion, and the Free Exercise Clause prohibits the state from inhibiting or restricting religion: taken together, the clauses require public institutions such as schools to be religiously neutral.

William Jennings Bryan had argued that neutrality consisted of teaching neither evolution nor creationism in the schools; antievolution laws removed evolution from the curriculum so that students would not be exposed to what some considered an antireligious doctrine. As evolution returned to textbooks and to the curriculum, creationists protested that the classroom was no longer neutral. To restore neutrality, they argued, *both* evolution and creationism should be taught.

A MOVEMENT BUILDS

The ICR encouraged citizens to take an active role in promoting creation science at the local level. In ICR's publication *Impact*, lawyer Wendell Bird encouraged local citizens to present school boards with resolutions encouraging the teaching of creation science in science curricula.[14] A conservative Christian layman, Paul Ellwanger, submitted his own resolution to the Anderson, South Carolina, school district, proposing a "balanced treatment of evolution and creation in all courses and library materials dealing in any way with the subject of origins."[15] Ellwanger subsequently prepared sample legislation for districts or states to pass, which was widely circulated.

Ellwanger's model legislation presented two alternative—and allegedly scientifically equivalent—views: "evolution science" and creation science, both of which he felt should be taught to maintain a balanced curriculum. If evolution was taught, schools would be required to teach creation science as well. Inspired by his efforts, a movement began to introduce Ellwanger bills in state legislatures. Although legislators in Ellwanger's home state of South Carolina failed to pass an Ellwanger bill, legislation soon began appearing in other states.

By the early 1980s, equal-time legislation had been introduced in at least twenty-seven states.[16] All died in committee, except for those in Arkansas and Louisiana. Many scientists and educators were involved in campaigns to prevent the passage of equal-time legislation. Creation science finally was receiving attention from scientists, though not the kind Henry Morris had desired.

MCLEAN V. ARKANSAS

The Ellwanger-inspired Arkansas Act 590 called for balance (by which it meant neutrality) between creationism and evolution but tried to avoid the Establishment Clause by claiming that creation science was *science*, rather than religion. Yet evolution was linked to religion because teaching evolution and not creationism was alleged to create a hostile climate for religious students. Teaching evolution only was held to be a violation of academic freedom, "because it denies students a choice between scientific models and instead indoctrinates them in evolution-science alone."[17] Creation science was presented as a strictly scientific view.

Upon its passage in 1981, the bill was swiftly challenged by the Arkansas American Civil Liberties Union. Plaintiffs in the lawsuit included Methodist clergyman William McLean and the bishops or other spokespersons for the Arkansas Episcopal Church, the United Methodists, Roman Catholics, African Methodist Episcopalians, Presbyterians, and Southern Baptists, as well as science education organizations, civil liberty organizations, and several individual parents. The presence of so many religious plaintiffs helped to defuse the argument that opposition to the bill equated to opposition to religion.

McLean v. Arkansas was tried in federal district court. Plaintiffs argued that because creation science was inherently a religious idea, its advocacy as required by Act 590 would violate the Establishment Clause. Furthermore, because creation science was not scientific, there was no secular purpose for its teaching. The state, defending the law, had to argue the opposite: that creation science was scientific, and thus its advocacy would have a secular purpose. Each side brought in expert witnesses to testify in favor of its position. Much time was spent in the trial over the definition of science, and whether creation science fulfilled it.

The judge ruled against the law: creation science was not science, and its teaching promoted a sectarian religious view. The judge noted that creation science proposed a "contrived dualism" referred to by proponents as the two-model approach: there are only two possibilities, special creationism or evolution. Hence, disproving evolution would prove creationism. Creation science, the judge noted, consisted largely of the presentation of arguments against evolution rather than a positive presentation supporting special creationism. It is a theme we will encounter again in the later creationist movement of intelligent design.

Equal time for creation science and evolution had failed in Arkansas, but another Ellwanger-derived law very similar to the Arkansas one had been introduced into neighboring Louisiana only a few months before the *McLean* decision.

THE LOUISIANA EQUAL-TIME LAW

The framers of the Louisiana law Balanced Treatment for Creation Science and Evolution Science in Public School Instruction, also passed in 1981, sought to avoid defining creation science in recognizably religious terms. Again the ACLU challenged the law in federal district court, but because proponents of the law also requested an injunction, courts had to sort out jurisdictional issues, and both cases slogged through the courts for several years. The district court tried the case by summary judgment: the judge accepted written statements from both sides and decided the outcome of the case based on these documents. Unlike *McLean,* then, the Louisiana case was not a full trial.

In 1985, the federal district court decided that the Louisiana law was unconstitutional because it advanced a religious view by prohibiting the teaching of evolution unless creationism—a religious view—was also taught. The court of appeals agreed, and finally, in 1987, the case made its way to the Supreme Court. The highest court concurred with the lower ones.

> *The preeminent purpose of the Louisiana Legislature was clearly to advance the religious viewpoint that a supernatural being created humankind. . . . The Louisiana Creationism Act advances a religious doctrine by requiring either the banishment of the theory of evolution from public school classrooms or the presentation of a religious viewpoint that rejects evolution in its entirety* (Edwards v. Aguillard, 482 U.S. 578 [1987]).

The ruling is explained in further detail in chapter 4.

Equal time for creation science was no longer a legal option in the schools of the United States. But as early as the defeat of creation science in *McLean*, a group of conservative Christians had begun searching for an alternative antievolution view that would not only be legally viable but would also appeal to a broader range of Christians. Creation science, with its stress on biblical literalism and the young Earth, attracts conservative Christians but to most mainstream Christians it appears to be marginal theology and odd science. This alternative became the intelligent design movement.

Intelligent Design

THE FOUNDATION FOR THOUGHT AND ETHICS

In 1980, Jon Buell, a former campus minister, established the Foundation for Thought and Ethics (FTE), which declared its primary purpose as "both religious and educational, which includes, but is not limited to, proclaiming, publishing, preaching, teaching, promoting, broadcasting, disseminating, and otherwise making known the Christian gospel and understanding of the Bible and the light it sheds on the academic and social issues of our day." FTE served as the nucleus for a group of conservative Christians concerned about the failure of creation science and the continued instruction of evolution. Accordingly, the publications and conferences sponsored by FTE and its allies focused on special creation's core idea of God the Designer, without direct reliance on the Bible: references to a universal Flood, to the special creation of Adam and Eve or any other creature, or to a young Earth were sparse.

Buell recruited the historian and chemist Charles Thaxton, the engineer Walter Bradley, and the geochemist Roger Olsen to write a document on scientific difficulties concerning the origin of life. This became FTE's first book, *The Mystery of Life's Origin*.[18] Paralleling the approach of creation science, *Mystery* emphasized supposed scientific shortcomings of scientific theory regarding the origin of life. Indeed, in 1984 as well as today, scientists have not reached consensus on the steps required for a natural origin of the first replicating structure; the origin of life is a currently unexplained event. But *Mystery* went beyond the observation that the origin of life is unexplained to conclude that the origin of life represented a category of scientific problems that was unexplain*able*. The origin of life, they claimed, could not be determined by natural laws alone. To explain the first cell, they invoked the action of an intelligent agent.

Since the publication of *Mystery*, new phenomena have been added to this list of putative unexplainable phenomena, including molecular machines such as the bacterial flagellum and complex biochemical processes such as the blood-clotting cascade and the immune system. Echoing the late eighteenth and early nineteenth century English clergyman William Paley, ID proponents claim that structural complexity cannot be explained by natural processes, which they equate with chance. (Chapter 2 describes their arguments in detail.) It is absurd to imagine chance forming complex structures; they must thus require the design of an intelligent agent.

The authors of *Mystery* did not identify the creative agent because, they said, they were examining the issue from a strictly scientific perspective, rather than a religious one. While admitting that they themselves believed the agent was the Christian God, they offered the alternative of Hoyle and Wickramasinghe[19] that life on Earth was produced by extraterrestrials of high intelligence. So the intelligent agent was not identified, although it would be difficult to avoid the conclusion that the agent was God.

More recent intelligent design proponents have not strayed from this claim, painstakingly attempting to distinguish ID from creationism. God might be the agent, but the agent might be material as well. It is a careful—if ultimately unsuccessful—dance between leaving in enough religion to keep creationists happy but keeping God at a sufficient distance to avoid the Establishment Clause.

OF PANDAS AND PEOPLE—AND CREATIONISM

The next book to emerge from FTE was the high school supplementary textbook *Of Pandas and People* [20] (*OPAP*). *OPAP* was the first book to identify itself with the phrase *intelligent design*. Its history illuminates the creationist roots of ID.

FTE began work on *OPAP* shortly after the *McLean* trial. In *Origins Research*, published by the group Students for Origins Research, it was reported, "A high school biology textbook is in the planning stages that will be sensitively written to 'present both evolution and creation while limiting discussion to scientific data.'" The authors of the textbook, Dean Kenyon and Percival Davis, both had deep creationist roots. Davis was the coauthor of a young-Earth creationist book, and Kenyon had written a foreword for *What Is Creation Science?*, by Henry Morris and Gary Parker, as well as an affidavit for the state of Arkansas defending Act 590. Unsurprisingly, the orientation and content of *OPAP* clearly reflects the authors' creationist roots.

Indeed, it turns out that *OPAP* was originally even more clearly rooted in creationism than the published version indicates. During the discovery phase of *Kitzmiller v. Dover*, the 2005 trial over a policy of the Dover, Pennsylvania, school board requiring the teaching of ID, early drafts of *OPAP* were subpoenaed by the plaintiffs' legal team and introduced during trial. In these drafts, titles and wording reflecting creationism morph into intelligent design. The earliest known draft is titled "Creation Biology" and is dated 1983. In 1986, the title was changed to the similar "Biology and Creation." A less creationist-sounding title was given 1987's "Biology and Origins" (although *origins* is a ubiquitous term in creationist literature and much less common in evolutionary biology), and in that same year, the first draft bearing the title "Of Pandas and People" is found. A second draft of this title is also dated 1987, and the first edition of the book was published in 1989.

Changes taking place in one paragraph in *OPAP* illustrate the creationist roots of ID. In the 1986 "Creation Biology" and the 1987 "Biology and Origins" manuscripts, creationism is defined as follows:

> Creation *means that the various forms of life began abruptly through the agency of an* intelligent creator *with their distinctive features already intact—fish with fins and scales, birds with feathers, beaks, and wings, etc. [Italics mine.]*

This, of course, is the definition of special creationism: the Christian doctrine that God created things in their present form. In 1987 the first manuscript with the title "Of Pandas and People" appears—with the identical wording of the earlier, overtly creationist manuscripts. This first OPAP-titled manuscript is something of a missing link, having the old special creationism wording yet using the new title. In a second 1987 *OPAP* manuscript the definition of creationism morphs into the definition of intelligent design:

> Intelligent design *means that the various forms of life began abruptly through an* intelligent agency, *with their distinctive features already intact—fish with fins and scales, birds with feathers, beaks, and wings, etc. [Italics mine.]*

The intelligent creator of the earlier versions also disappears in this second 1987 manuscript (and in the first and second editions of the published version of the book) to be replaced by the less religious-sounding phrase *intelligent agency*. Why did the wording change between these two versions of the 1987 manuscripts, reducing the overtly creationist language? It is perhaps not coincidental that the *Edwards v. Aguillard* decision was announced in July 1987, making creation science a legally invalid strategy.

The 1989 OPAP first edition was the first time that the term *intelligent design* appeared in print to describe the form of creationism expressed in *The Mystery of Life's Origin*. As OPAP's history clearly reflects creationist roots, so also does the content of ID. Creation science founding father Henry Morris in fact chided ID theoretician William Dembski for not giving sufficient credit to creation science predecessors:

> *These well-meaning folks did not really invent the idea of intelligent design, of course. Dembski often refers, for example, to the bacterial flagellum as a strong evidence for design (and indeed it is); but one of our ICR scientists (the late Dr. Dick Bliss) was using this example in his talks on creation a generation ago. And what about our monographs on the monarch butterfly, the bombardier beetle, and many other testimonies to divine design? Creationists have been documenting design for many years, going back to Paley's watchmaker and beyond.*[21]

In both creation science and ID, a familiar litany of topics can readily be found, all of which highlight the supposed inability of evolution to explain the origin and diversity of living things. The absence of a scenario for a natural origin of life, of course, is a mainstay of both creation science and ID: it was the topic of the founding document of ID, *The Mystery of Life's Origin*. The supposed lack of transitional fossils ("gaps in the fossil record") is a mainstay of both creation science sources such as Duane Gish's *Evolution: The Fossils Still Say No!*[22] and ID sources such as Jonathan Wells's *Icons of Evolution*.[23] Both ID and creation science proponents view the Cambrian Explosion of invertebrate body plans as something impossible to explain through evolution.[24] Another familiar topic in the literature of the two forms of creationism is the supposed inability of natural selection to account for complex biological structures: it is the key idea of ID, but it was presaged in creation science writings for decades. Also present in both creation science and ID is an effort to refute evolution through probability arguments: Henry Morris frequently cited the impossibility of assembling even a simple protein through chance, and William Dembski's *The Design Inference*[25] is the key ID text for validating design—through probability.

Other themes common to both creation science and ID include a mistrust of naturalism, both as a methodology of science (because it restricts science to natural causes) and as a philosophy (because it claims that only natural causes exist). Limited evolution is recognized, but only within strict limits; in creation science, this is referred to as evolution within the kind.[26] And—of course—Christianity permeates the literature of both groups, and proponents view both movements as ministries for bringing the unsaved to Christ.[27]

The efforts of FTE to promote ID through *OPAP* did not generate much public notice, perhaps because FTE as a Christian ministry did not attract much notice from the mainstream press. The ID movement became much more widely known with the publication of a book by law professor Phillip Johnson.[28] The advent of Johnson on the ID stage appreciably changed the movement.

PHILLIP JOHNSON AND INTELLIGENT DESIGN

While on a sabbatical at Cambridge University in 1987, Johnson met an American philosophy student named Stephen Meyer who introduced him to the neo-creationist antievolutionists associated with FTE. Upon returning to the University of California at Berkeley's law school, Johnson began participating in design-oriented conferences and writing *Darwin on Trial*. The publication of an antievolution book by a tenured professor at a major secular university such as Berkeley came as a surprise to the educated public. Although books by Henry Morris were ignored by the scientific community, a few scientists reviewed *Darwin on Trial* in popular publications such as *Scientific American,* and discussions of this new form of antievolutionism appeared in the popular press. Still, scientists uniformly criticized what they considered to be uninformed science in Johnson's book. Johnson's subsequent books focused more on social, educational, and theological issues, but they continued expressing the basic ID message that biological structural complexity was impossible to achieve through unguided natural causes; it required God's direct action. Johnson also forcefully expressed the ID animosity toward philosophical naturalism—a belief that reality consists only of material phenomena (matter and energy and their interactions) and that there is no supernatural. He also criticized the methodological naturalism of science—the well-established practice of limiting science to only natural causes. This, too, is a theme in creation science, but it became central to ID, which proposed as early as *Of Pandas and People* (second edition) that science should be expanded to include the occasional supernatural intervention.

Johnson's major contribution to the ID movement was strategic. He argued for "mere creation" (the unifying of all antievolutionists around the uniformly accepted concept of design) and urged putting aside issues such as biblical literalism, the age of the Earth, Flood geology, and other traditional creationist themes. His view was that "people of differing theological views should learn who's close to them, form alliances and put aside divisive issues 'til later. I say after we've settled the issue of a Creator, we'll have a wonderful time arguing about the age of the Earth."[29] Perhaps because of Johnson's importance and influence on the movement, attention began to shift away from Jon Buell and FTE in the mid-1990s. In 1996, leadership of the ID movement passed to the Discovery Institute think tank in Seattle, Washington.

ID GROWS: THE DISCOVERY INSTITUTE AND THE WEDGE

The Discovery Institute is a think tank founded in 1990 by a former politician named Bruce Chapman. The DI is an umbrella organization that houses a number of centers or projects dealing with such issues as regional transportation, technology, economics (mostly tax and free-market policy), religion and public life, and law and justice (tort reform). The intelligent design–promoting Center for Renewal of Science and Culture (CRSC), headed by Stephen Meyer, was announced in a 1996 press release:

> *For over a century, Western science has been influenced by the idea that God is either dead or irrelevant. Two foundations recently awarded Discovery Institute nearly a million dollars in grants to examine and confront this materialistic bias in science, law, and the humanities. The grants will be used to establish the Center for the Renewal of Science and Culture at Discovery, which will award research fellowships to scholars, hold conferences, and disseminate research findings among opinion makers and the general public.*[30]

Later, the name of the unit was changed to the less religious-sounding Center for Science and Culture (CSC). Since 1996, the Discovery Institute's intelligent design unit has sponsored the writing of books by postdoctoral scholars, the production of antievolution videos, and an extensive media campaign including a frequently updated Web site. This campaign includes frequent submission of op-eds and press releases and an ongoing evaluation of the coverage of ID in the media—most of the latter consisting of complaining about the alleged unfairness of any coverage that is not uniformly positive.

The Web site touts the scientific research and scholarship of ID, presenting a list of peer-reviewed articles and peer-edited books and conference proceedings as examples of this new science. The list is paltry, and most of the items on it would not generally be regarded as contributions to the scientific research literature; one item, in fact, was repudiated by the journal in which it appeared. But the more important point is that it shows that nobody is using ID to advance our knowledge about the natural world. Although the scientific claims of ID will be examined in more detail in chapter 2, let me take a moment and briefly describe what the central scientific claim of ID turns out to be. It will become clear why ID is scientifically sterile.

ID'S SCHOLARLY PRETENSIONS

William Dembski defines ID as composed of three parts: "A scientific research program that investigates the effects of intelligent causes; an intellectual movement that challenges Darwinism and its naturalistic legacy; and a way of understanding divine action."[31]

Although Dembski gives primacy to scholarship, in actuality, ID's scholarly pretensions are thin. There are only two concepts comprising ID theory: Michael Behe's "irreducible complexity" and William Dembski's "complex specified information," which is related to what he calls the design inference. Both are attempts to identify those phenomena in nature that supposedly are unexplainable through natural causes—the structurally complex systems and molecular machines that require the action of the intelligent agent. As will be discussed in chapter 2, both irreducible complexity and the design inference have been rejected by scientists as being irrelevant to the understanding of biology.

In content, ID resembles the traditional two-model approach of creation science. As in creation science, the underlying assumption is that either evolution or an intelligent agency explains nature. Disproving evolution leaves ID as the default winner. As a result of this mindset, ID literature focuses on problems with evolution. Jonathan Wells's *Icons of Evolution* is a classic in this genre, presenting generally misleading and/or inaccurate information on topics such as the peppered moth example of natural selection, the Miller-Urey sparking experiments producing amino acids, homology, the Cambrian Explosion, and human and bird evolution. Evolution is presented as having insurmountable flaws, with the default solution to this problem being the presumption of a Designer. The critiques of evolution offered in such ID literature, however, is recognizable as a proper subset of the critiques offered by creation science literature, and they are no more valid.

TAKING ID ON THE ROAD

Especially in the early days of the CSC, ID was promoted through a series of conferences, the goals of which were to introduce ID to both the public and the scholarly community and to increase acceptance of the basic ID premise that evolution was inadequate science and that ID presented a fresh and valid perspective.

From my own observations and from interviewing scientists who had attended similar ID programs, it appears that the organization of ID conferences during the late 1990s and early 2000s followed a pattern. The official approach—on stage at least—was secular; for example, the program did not open with prayer, and references to religion were minimal. The ID proponents presented their science, the critics pointed out flaws, the ID proponents thanked the critics for their openness in engaging in scholarly dialogue, and they proceeded to the next conference at which exactly the same arguments on both sides were again presented, but to a fresh audience of laypeople and academics.

Eventually, the Discovery Institute stopped asking critics to attend these conferences, and a second conference format began to evolve. At these second-generation conferences, only ID proponents presented, the audience was largely sympathetic, and there were more overt religious references. At a 2002 conference organized by the campus-based student Intelligent Design and Evolution Awareness (IDEA) club in San Francisco, for example, one of the breakout speakers began his session with a prayer. Keynote speakers referred to apologetics, and the audience seemed largely recruited from churches and local religious colleges.

ID advocates complain that their views are rejected out of hand by the scientific establishment, yet they do not play by the normal rules of presenting their views first through scientific conferences and then to peer-reviewed journals and then in textbooks. Few ID proponents present papers or posters at professional meetings, and those who do attend such meetings do not present examples of the use of central ID concepts such as irreducible complexity and the design inference. Instead, they present vaguely antievolutionary papers or papers discussing the sociological aspects of the controversy over ID and evolution. And significantly, the first publication to use the phrase *intelligent design* was not a theoretical paper but a high school textbook, *Of Pandas and People*! Ordinarily, one does the research first and *then* produces the textbook.

Now, instead of attempting to persuade the scholarly community, ID proponents bypass scientists and go directly to the general public, where they have been much more successful. Discovery Institute fellows such as Phillip Johnson, Steven Meyer, Michael Behe, Jonathan Wells, and Paul Nelson have been able to publish ID views in op-eds in national publications such as the *Wall Street Journal,* the *New York Times,* and the *Washington Post,* and in major regional outlets such as the *Los Angeles Times*. Their success in swaying public opinion is evidenced by the large number of school boards that have considered introducing ID into their curricula and state legislators who have introduced pro-ID legislation over the last few years.

They did this by staying on message that ID was not creationism but a new form of science, and by attacking evolution as inadequate science. Ironically, to anyone familiar with the history of the antievolution movement, the attacks on evolution are perhaps the most obvious link between ID and earlier forms of creationism.

THE WEDGE

The most complete single source of information on the goals and strategy of the ID movement is a fundraising proposal to an unknown foundation or individual prepared by the Discovery Institute in the late 1990s, called "The Wedge Strategy." The manuscript of the fundraising proposal was leaked, posted on the Internet, and eventually acknowledged by the Discovery Institute. The Wedge document, as analyzed by Barbara Forrest and Paul R. Gross,[32] lays out general goals and five-year and twenty-year plans. The overall goals of the ID movement are explicitly religious:

> *To defeat scientific materialism and its destructive moral, cultural and political legacies. To replace materialistic explanations with the theistic understanding that nature and human beings are created by God.*

ID is revealed to be a specific sectarian, Christian view:

> *Design theory promises to reverse the stifling dominance of the materialist world view, and to replace it with a science consonant with Christian and theistic convictions.*

The five-year goals of ID begin with plans to establish ID as a valid science but extend further:

> *To see intelligent design theory as an accepted alternative in the sciences and scientific research being done from the perspective of design theory. To see the beginning of the influence of design theory in spheres other than natural science.*
>
> *To see major new debates in education, life issues, legal and personal responsibility pushed to the front of the national agenda.*

ID, like creation science, has goals that are primarily religious. And, like creation science, ID has not been fully successful in hiding its fundamentalist motivations, which its proponents have candidly divulged when speaking to the conservative Christians who form the base, or ground troops, of the movement.

THE FUTURE

We have seen that ID is a subset of the ideas presented in creation science: it focuses narrowly on the big idea that God created, and it sets aside the details of creation science such as six-day creation, a young Earth, Flood geology, and the like. It retains the core dichotomy of creation science, which is that evidence against evolution equates with evidence for creationism. It is only necessary to disprove evolution and creationism will then be proved by default. Promotion of this two-model approach is in fact the direction ID is heading.

"TEACH THE CONTROVERSY"

Originally, the DI argued that intelligent design deserved a place not only at the academic table but also in the public schools. In the late 1990s the CSC published white papers and op-eds arguing for the academic appropriateness and legality of teaching ID in high school.[33] If ID was banned, a district would be guilty of "viewpoint discrimination," contended the ID legal experts.

But in the early 2000s the DI proponents of intelligent design seemingly had a change of heart. Rather than lobbying for the teaching of ID, they shifted tactics to promoting the teaching of straight antievolutionism. This was not a huge change in content, of course, as ID had always stressed the weaknesses or flaws of evolution, and in fact, ID is scarcely anything *other* than pronouncements about the weakness of evolution as a scientific theory. For ID advocates, the new approach was merely a change in emphasis. There are a number of phrases employed to promote this approach, as is discussed further in chapter 5. Teachers are variously exhorted to "critically analyze" evolution, to teach the "strengths and weaknesses" of evolution, to teach "evidence for and evidence against" evolution, to teach "the full range of views" about evolution, to teach "evolution as theory not fact," and of course, to "teach the controversy."

Although those outside the halls of the Discovery Institute can only speculate as to why this change in emphasis was made, it may be that advocates realized the inherent weakness of the phrase *intelligent design*, which implies a designer. A judge seeking the identity of the intelligent agent would quickly conclude—even from statements from Discovery Institute fellows themselves—that the agent is the Agent, exposing ID to the same legal liability that creation science experienced. The DI may have believed that proposing schools teach that evolution is flawed or weak doesn't appear to be promotion of religion. Teaching bad science is perhaps unwise educational policy, but it might not be unconstitutional. It might also have occurred to the ID proponents that because there is no curriculum for ID other than "evolution doesn't work" (as in the *Icons of Evolution* arguments and examples), the argument that ID should be introduced into the curriculum is weak.

But perhaps the best argument for shifting from advocacy of ID to advocacy of "teaching the controversy" (teaching antievolutionism) is that it appeals to American sensibilities of fairness: who wouldn't want students to be exposed to all the evidence, to the full range of scientific views, and to be able to make up their own minds, thus improving their critical thinking skills?

When the Dover, Pennsylvania, school district passed a policy requiring the teaching of ID, the Discovery Institute opposed it, favoring a "teach the controversy" approach instead. After the 2005 *Kitzmiller v. Dover* trial ended in a rout for the scientific pretensions of ID, the Discovery Institute attempted to attack the decision, defending the scientific plausibility of ID while simultaneously denying that ID was ready for the classroom. Given the strength of the *Kitzmiller* decision, it is doubtful that the DI will promote any further policies encouraging the teaching of ID. (The details of the *Kitzmiller* decision are discussed in detail in chapter 4.)

We can anticipate that ID advocates nonetheless will promote policies at the school board and state legislative levels that inhibit the teaching of evolution in some fashion. This could be through requiring evolution disclaimers (although such disclaimers have twice been found to be unconstitutional in federal courts), policies requiring that evolution be taught as theory, not fact (with *theory* here understood in the vernacular sense of a hunch or guess), or policies directing teachers to teach the erroneous scientific content on evolution encountered in ID literature. An approach that is appearing in the middle of the first decade of the twenty-first century is to offer permissive policies and/or legislation that allows teachers to present ID or antievolutionism without fear of being punished. These academic freedom bills are intended to encourage teachers who wish to bring creationism into the curriculum to do so without fear of legal sanction. None of these latter approaches yet has been tested in court.

Antievolutionists are realizing that teaching creation science or intelligent design may be superfluous if teaching straight antievolutionism will do the job of discrediting evolution in favor of a religious view. And this, of course, has been and remains the goal of the antievolution movement, beginning with the Scopes-era focus on banning evolution. Evolution is viewed as atheistic, and religious conservatives believe that its acceptance by students will lead them to abandon their faith. The two-model approach is alive and well: if evolution is disproved, students will naturally default to special creationism.

This is the message of creation science and of intelligent design. It will continue to be the message of whatever evolves from intelligent design in the future.

2: Analyzing Critical Analysis: The Fallback Antievolutionist Strategy

NICHOLAS J. MATZKE AND PAUL R. GROSS

Introduction

The primary challenge to teaching evolution in public school science classrooms no longer comes labeled as *creationism, creation science,* or even *intelligent design*. The strategies those labels represent have all failed in the federal courts because they boil down to special creation—the theological doctrine that God intervened miraculously in the history of life. Plainly, this is a specific religious claim, not even shared by all Christian denominations. Furthermore, because the U.S. Constitution prohibits an establishment of religion by government, it prohibits government-employed science teachers from presenting religious views as though they were, or had the support of, genuine science.

Creationism does not become extinct; it evolves. The history of creationism (as reviewed in chapter 1) makes it clear that court defeats cause creationists to change labels and legal strategies but do little or nothing to change their underlying objective. In fact, by the time a definitive court case settles the fate of one tactic, legal difficulties have usually become apparent, and another tactic is already in the works. Creation science was developed in the years preceding the Supreme Court's 1968 *Epperson* decision, which overturned bans on teaching evolution; intelligent design developed between the crushing defeat of creation science in the 1982 district court decision *McLean v. Arkansas* and the 1987 Supreme Court *Ed-*

wards decision. (These cases are discussed further in chapter 4.) Given the defeat of intelligent design in *Kitzmiller v. Dover* in 2005, what can we expect to see next by way of creationist attacks?[1]

It appears certain that the main challenge to teaching evolution in public schools will be educational policies that propose critical analysis and similar invocations of critical thinking—specifically in connection with evolution-related science. Reviewing the history and origins of these critical analysis interventions, and examining the common so-called evidences against evolution that creationists have promoted in the guise of critical analysis—for example in Kansas, Ohio, and South Carolina—it becomes clear that while labels have changed, the content of these standards and their accessory material remains the same. All the critical analysis arguments are traceable to primary texts of the intelligent design (ID) and creation science (CS) movements. They are, without exception, aimed at promoting the sectarian doctrine of special creation.

Origins of the Critical Analysis Tactic

The critical analysis tactic has deep roots in the creationist movement. In the face of political and especially legal opposition to policies requiring equal time for creation science, creationists in the 1980s suggested alternatives. For example, following the *Edwards* decision against creation science in 1987, the Institute for Creation Research hoped for another major test case that would reach a different result but in the interim recommended a new tactic:

> *In the meantime, school boards and teachers should be strongly encouraged at least to stress the scientific evidences and arguments* against evolution *in their classes (not just arguments against some proposed evolutionary mechanism, but against evolution* per se*), even if they don't wish to recognize these as evidences and arguments* for creation *(not necessarily as arguments for a particular date of creation, but for creation* per se*). To do anything less is equivalent to making humanistic evolution an article of faith, and this would be an establishment of religion!*[2]

As for the ID movement, it was clear even before *Kitzmiller* that it was sidling away from policies that encouraged intelligent design directly. In 1998, the Discovery Institute's Center for the Renewal of Science and Culture was clearly committed to teaching ID in the public schools. For example, the Discovery Institute's strategic plan, the Wedge Strategy, had the following as one of its five-year goals: "Ten states begin to rectify ideological imbalance in their science curricula and include design theory."[3] This is an unambiguous endorsement of efforts to make ID a requirement of curriculum. By 2004, however, the Discovery Institute was denying adamantly that it had ever advocated such a thing. Instead, it was promoting critical analysis of evolution, teaching the strengths and weaknesses of evolution, or teaching the controversy.

The beginning of the ID movement's shift toward critical analysis can be traced to two factors. The first is *Icons of Evolution* (2000), a blustering antievolution polemic written by Discovery Institute fellow Jonathan Wells, which succeeded in intimidating many educators by attacking biology textbooks for allegedly crude mistakes or worse.[4] The second is the so-called Santorum amendment, a nonbinding "sense of the Senate" provision that Pennsylvania senator Rick Santorum attempted to insert into the No Child Left Behind Act in June 2001. Santorum's amendment (written, in fact, by ID leader Phillip Johnson) did not mention creationism or ID, but it did single out evolution as controversial science. The amendment was eventually stripped from the statute language. Instead, a watered-down version appeared in the joint explanatory statement of the House-Senate conference committee. This statement had no force as law, but it was nevertheless trumpeted by antievolutionists as a signal victory for their cause.

The Santorum language was cited in subsequent battles over critical analysis policies, and it paved the way for the ID movement's two major state-level victories: critical analysis language in the science standards of Ohio (2002) and Kansas (2005). Creationists and ID advocates have also pushed unsuccessfully for critical analysis language in the science standards in Minnesota, New Mexico, Pennsylvania, South Carolina, West Virginia, Arizona, and Georgia. Of the proposed antievolution legislation tracked by the National Center for Science Education in 2005, many of the bills employed critical analysis or similar euphemisms for the deprecation of evolutionary science. Furthermore, many local school boards have been the scenes of battles over similar policies—prominent examples include Roseville, California; Darby, Montana; Grantsburg, Wisconsin; and Cobb County, Georgia.[5]

OHIO, 2002

Critical analysis of evolution was codified as the ID movement's favorite tactic by its insertion into the Ohio science standards in 2002. The federal No Child Left Behind Act was passed in December 2001, and the Santorum language was immediately put to use in a science standards battle brewing in Ohio. In the wake of proposals to include ID in the Ohio standards, the Ohio Board of Education scheduled a debate between proponents of evolution (Kenneth Miller of Brown University and Lawrence Krauss of Case Western Reserve University) and advocates for intelligent design (Discovery Institute senior fellow Jonathan Wells and Stephen C. Meyer, director of the Discovery Institute's ID program). In a surprise move, Meyer announced that he did not want ID taught; he offered, instead, a compromise, which was simply to critically analyze evolution. The board eventually adopted Meyer's position. The resulting long (and generally thoughtful and correct) standard concerning evolution read, in one place, "Describe how scientists continue to investigate and critically analyze aspects of evolutionary theory." In response to criticism of this, a parenthetical qualifier was added: "The intent of this indicator does not mandate the teaching or testing of intelligent design." But the Discovery Institute and the Intelligent Design Network (a Kansas-based ID-advocacy organization) saw the final result as a great victory.[6]

They had reason to be encouraged by what happened after the Ohio Board of Education passed the otherwise acceptable science standards. The time came to write model lesson plans based on the standards. Predictably, creationists got themselves on the committee writing the lesson plan for standard L10H23, Critical Analysis of Evolution. The lesson produced was clearly based on the creationist/ID literature, especially *Icons of Evolution*. The lesson targeted common descent. The original title for the lesson plan was The Great Macroevolution Debate.[7] Under criticism, that sample lesson was cut back substantially—the title was changed, some of the *Icons*-derived material was dropped, and references to nonexistent articles and to creationist and Christian apologetics Web sites were deleted. The resulting lesson was passed in 2004—but after a national review of state standards awarded a grade of only B to Ohio's massive, complicated, but generally competent science standards.[8] The lesson included standard creationist arguments about homology, the fossil record, antibiotic resistance, peppered moths, and endosymbiosis, and it misled students about a distinction between microevolution and macroevolution. Because its use was, in principle, optional for teachers, it was not officially a standard or a benchmark; this made it difficult to challenge legally.

Supporters of the lesson nevertheless stated publicly that they had developed a lawsuit-proof antievolution tactic. The legal question was mooted in February 2006, when in the wake of the *Kitzmiller* decision and a series of embarrassing revelations from Freedom of Information Act requests, the Ohio Board of Education reversed itself and voted 11–4 to eliminate the Critical Analysis of Evolution lesson plan and the language in the standards documentation on which it was based. Up to the last moment, however, it was a close call.

KANSAS, 2005

The critical analysis tactic achieved its most prominent success to date in the 2005 Kansas science standards. The Kansas-based Intelligent Design Network led the charge here. But despite its name, it did not push for an explicit intelligent design requirement in the Kansas science standards revision of 2005. This was because of the national developments already described and because the creationists on the Kansas Board of Education had been voted out of office in 2000 after deleting evolution and related concepts from the standards in 1999. (The new board restored evolution to the standards in 2001.) Instead, when creationists again won a majority in the 2004 elections, the Intelligent Design Network persuaded the board to include "scientific criticisms of [evolutionary] theory," encouraging students thereby to "critically analyze the conclusions that scientists make."[9]

The board's official rationale stands as a monument to the doublespeak of this tactic. The critical analysis move was justified by this assertion: "The Board has heard credible scientific testimony that indeed there are significant debates about the evidence for key aspects of chemical and biological evolutionary theory." It failed to mention that said scientific testimony was just the set of presentations from some twenty creationists and intelligent design proponents the board itself had invited to speak in May 2005. Almost all of those speakers denied a common ancestry of humans and apes, and some stated that they thought the Earth was only thousands of years old.[10]

Left unsaid in the rationale was the fact that all the Intelligent Design Network's changes to the science standards were taken directly from creationist/intelligent design literature. Nevertheless, in keeping with the critical analysis tactic, the board "emphasize[d] that the Science Curriculum Standards do not include Intelligent Design," even though (in the very same sentence!) the board promoted the scientific credibility of ID, describing it as "the scientific disagreement with the claim of many evolutionary biologists that the apparent design of living systems is an illusion." The idea that living things exhibit design is not foreign to biology, contrary to the board's suggestion, but no competent biologist thinks that the presence of design in living things necessarily implies a conscious, purposeful designer.

The board concluded that "these standards neither mandate *nor prohibit* teaching about this scientific disagreement" (emphasis added), thus giving a wink and a nod to any creationists on local school boards who might be paying attention. The numerous edits to the science standards included a number of favorite creationism/ID talking points. In effect, the Intelligent Design Network, by way of the Kansas Board of Education, put a government-sponsored advertisement for ID on the front page of the Kansas science standards, and inserted everything except the actual words *intelligent design* into the science benchmarks. The attempt is to avoid legal challenge by *encouraging* rather than *requiring* ID, and by cloaking the standard ID talking points in such locutions as "critical thinking," teaching "the full range of scientific views," and "scientific criticism."

CRITICAL ANALYSIS OF EVOLUTION AS A POLITICAL TACTIC

Like creation science and intelligent design, critical analysis is not a coherent scholarly undertaking but rather a political *tactic*. In all three approaches, the goal is to appear to make scientific rather than religious arguments, with the hope that science teachers, the public, and especially the courts will perceive this as legitimate government activity rather than as governmental promotion of a particular religious view. Because it is impossible to gather physical scientific evidence to support any hypothesis of supernatural causation, CS and ID attempted to establish special creation primarily through arguments against evolution. Relying on the contrived dualism that evolution and creationism are the only two possible explanations of life's origins and history, the CS and ID movements try to disparage evolution and then argue that special creation is the only alternative. Thus, arguments against evolution comprise the vast bulk of all the CS/ID literature. ID is just a subset of the claims made by CS, and critical analysis, deployed as described here, is just a subset of ID.

Critical Analysis of "Critical Analysis"

Shouldn't we encourage truly critical thought about dominant scientific theories? The answer, of course, is yes. But that is not what these policies are doing. They are, instead, uncritically promoting creationism-driven pseudoscience, in full accord with the traditional creationist goal: getting public school science classes to teach as science the theological doctrine that God intervened to create each kind of organism, including humankind. This is effected by singling out evolution *alone* for special criticism, ignoring all other major scientific theories, including those about which there really is a current argument (the relation between quantum mechanics and gravity, for example). To complete the job, long-debunked creationist criticisms of evolution are presented as though they are real, current science.

The problem for science education is that critical analysis policies promote creationist pseudoscience, or just bad science known to be fallacious. Students will simply be misinformed if these claims are taught as if they are accepted science. Supporters of first-rate science education need to be aware of critical analysis claims and some of the problems with these claims, because they are often disguised in science-style phrases and promoted as cutting-edge science. To illustrate, we offer here some actual critical analysis of the "critical analysis" items presented in the 2005 Kansas science standards.

INCONSISTENCIES IN PHYLOGENETIC TREES

The new creationism offers a standard list of challenges to the evidence that modern organisms share common ancestry. The Kansas science education standards that were adopted on November 8, 2005, list several. The first deals with disagreements between phylogenetic trees (lineages, or patterns of descent) constructed from DNA or protein-similarity analyses:

> *The view that living things in all the major kingdoms are modified descendants of a common ancestor (described in the pattern of a branching tree) has been challenged in recent years by:*
> *i. Discrepancies in the molecular evidence (e.g., differences in relatedness inferred from sequence studies of different proteins) previously thought to support that view.*

The claim is that phylogenetic trees based on different data sets conflict so badly as to call common ancestry into question. The usual creationist procedure is to dig through the scientific literature to find cases where studies disagree on the *exact* phylogenetic relationships of organisms and then to trumpet these as inexplicable discrepancies that refute common ancestry. ID creationists universally fail to acknowledge that the similarity of phylogenetic trees can be measured statistically, and that trees derived from independent data sets typically have extremely *strong* statistical correlations. Such findings, which are very common indeed, support the notion that there are real phylogenetic trees, and that scientists are mapping them. The touted disagreements, measured quantitatively, are rather like the disagreement between two independent dating methods for the age of the Earth, one giving the age as 4.50 billion years, and another giving it as 4.55 billion years—very similar measurements with a small amount of experimental error. Even phylogenies derived independently from morphological (anatomical) and molecular (chemical) data sets typically show a high degree of correlation.[11] Any ID/creationist claim that phylogenetic trees show discrepancies is worthless unless they report proper similarity statistics, and this they have never done. By contrast, a recent striking example of molecular and morphological (fossil) data coming into astonishing agreement is the documentation of a connection between the ancestors of whales and those of hippos.[12]

ABSENCE OF TRANSITIONAL FORMS AND THE CAMBRIAN EXPLOSION

The Kansas science standards state,

> *Patterns of diversification and extinction of organisms are documented in the fossil record. Evidence also indicates that simple, bacteria-like life may have existed billions of years ago. However, in many cases the fossil record is not consistent with gradual, unbroken sequences postulated by biological evolution.*

> *[Common ancestry is challenged by a] fossil record that shows sudden bursts of increased complexity (the Cambrian Explosion), long periods of stasis and the absence of abundant transitional forms rather than steady gradual increases in complexity.*

These two elements in the Kansas science standards make assertions about gaps in the fossil record. The ubiquitous creationist argument about transitional fossils is generally conducted by misquoting or distorting the words of a paleontologist, such as Stephen Jay Gould. Gould expressed outrage over this, memorably:

> *Since we proposed punctuated equilibria to explain trends, it is infuriating to be quoted again and again by creationists—whether through design or stupidity, I do not know—as admitting that the fossil record includes no transitional forms. Transitional forms are generally lacking at the species level, but they are abundant between larger groups.*[13]

Here, Gould refers to the Eldredge-Gould theory of Punctuated Equilibria, which simply took a standard model of speciation called allopatric speciation and applied it to the fossil record. In allopatric speciation, a small subpopulation of a species becomes geographically isolated. For reasons of population genetics, small populations are more likely to evolve quickly than large ones. Eldredge and Gould argued that if this were the dominant mode of speciation, then the fossil record should record relatively few smooth transitions between very closely related species. Gould was not saying that fossil species bridging large transitions, for example between mammals and their reptilelike ancestors, should be rare. *And they aren't*. In general, it is simply false to assert that transitional fossils (a century ago they were called missing links) are unknown or very rare. As Gould states, such transitions are abundant.

The second standard refers specifically to the Cambrian Explosion. Creationists consider the Cambrian Era as displaying the biggest gaps. For creationists, the Cambrian Explosion refers to the sudden appearance of complex animals—marine invertebrates—in the fossil record about 520 million years ago. In many cases, the first fossil record of a phylum occurs in the Cambrian—for example, the classic Cambrian animal is the trilobite, which belongs to the phylum Arthropoda along with insects, crustaceans, and other groups. Although the definition of *phylum* is somewhat arbitrary, other often-recognized phyla are the chordates (including vertebrates), mollusks (such as clams, squid, and octopuses), and echinoderms (starfish and sea urchins). In Darwin's time, the fossil record of these animals did appear to begin abruptly in the Cambrian, with no precursors, and creationists have attempted to exploit this gap ever since. The Cambrian Explosion is thus a prominent feature of all varieties of creationism and ID, and in several of the critical analysis proposals.

Unfortunately for the creationists, the Cambrian Explosion was not literally a sudden event: its duration was, conservatively, ten million years. But the most important fact that creationists ignore is the evidence from trace fossils and from recently discovered ancestral forms. While body fossils of soft-bodied organisms are very scarce, their crawling and burrowing activity disturbs sediments, leaving traces such as tracks and burrows that petrify and become fossils. This leaves a massive record in sedimentary rocks that must be taken into account. Before animals evolved, there was nothing macroscopic able to crawl around and eat the algal mats on the sea floor; therefore, such sediments are undisturbed. A few tens of millions of years before the animal phyla appear, however, worm tracks begin to appear—very simple, and then gradually increasing in complexity. At first, the worms moved only horizontally on the surfaces of the algal mats, but eventually they started to burrow into the sediment. Around the world, this bioturbation of the ocean floor meant an end to the world of undisturbed algal mats and the delicate Ediacaran frond forms: worms were eating everything! Just before the base of the Cambrian, about 543 million years ago, the first tiny shell is observed in the fossil record—*Cloudina*.

The shell was probably secreted to protect this worm from other worms that might eat it—evidence of shell borers is also preserved. With the start of the Cambrian, small, shelly fossils increase in diversity and complexity until classic Cambrian Explosion representatives like trilobites are observed—for example, in the fossil beds of Chengjiang, China, about 515 million years ago, and in the Burgess Shale around 500 million years ago. Creationists focus on these two fossil locations, but they completely ignore the earlier events recorded in the trace fossils and by the small shelled forms. When this full record is considered, the evidence indicates that in the early Cambrian, evolution was proceeding rapidly and many ecological niches were being occupied for the first time.[14]

As for transitional forms, recent discoveries are filling in the gaps even among the Cambrian phyla. Remember that the big picture painted by the trace fossils is of the gradual diversification of worms—relatively simple tubes with mouths. It should also be remembered that at least half of the animal phyla living today are *still* wormlike. Even for the various advanced phyla—mollusks, chordates, echinoderms, and arthropods—basal members of the phyla, or closely related sister phyla, are basically wormlike. So it is not surprising that paleontologists have discovered wormlike fossils in the Cambrian that share some, but not all, of the features of the modern phyla. For example, a diverse group of lobopod fossils illustrates the step-by-step acquisition of characteristics that define the living arthropods. Graham Budd, a specialist on the Cambrian arthropods, describes the real situation:

> *A remarkably complete series is now available, demonstrating how the most basal, worm-like taxa of the entire Arthropoda* sequentially acquired the important features characteristic of their clade... *Clearly, for the arthropods at least, current opinion now stands rather far away from the view expressed only a decade ago that the Cambrian record did not reveal anything of the origin of the phyla.*[15]

Simon Conway Morris, another leading specialist of the Cambrian, places a sluglike fossil named *Halkeria* as intermediate between the mollusk, brachiopod, and annelid phyla. As for chordates and echinoderms, early chordate relatives are represented by fossils like *Haikouella, Yunnanozoon,* and *Pikaia,* and early vertebrates by fossils like *Haikouichthys*—all small wormlike forms very different from what anyone would consider modern fish—and early echinoderm relatives include homalozoans and the remarkable vetulocystids.[16]

In short, evolutionary biologists have found abundant fossils with transitional characteristics showing how the body plans defining modern phyla were acquired step by step in the course of the Cambrian Explosion. Many contemporary leaders in the field have noted that the old issues—such as the contention that phylum-level characteristics are special and appeared only during the Cambrian Explosion, that stasis followed, or that the gaps among phyla were especially large—were based on the misapplication of the old Linnaean classification system to the fossil record, rather than on actual data. According to Simon Conway Morris, "the strangeness of the problematic Cambrian animals is really a human artifact, a construct of our imagination."[17] Creationists still cling to these old, discredited, and now imaginary problems, and have enshrined them in the Kansas science standards!

HAECKEL'S EMBRYO DRAWINGS

The Kansas science standards state that common ancestry is called into doubt by "studies that show animals follow different rather than identical early stages of embryological development." This is a key claim from Jonathan Wells's book *Icons of Evolution*. The argument is this: evolution is demonstrated by the embryological similarities shown in Ernst Haeckel's famous embryo drawings, but Haeckel faked these drawings to make the embryos more similar than they actually were, and this fake evidence for evolution has been reproduced in textbooks for school use.

The facts: Haeckel did exaggerate similarities in very early embryos of different species, and his figures, or derivatives of them, have appeared in a few textbooks (three of the ten textbooks that Wells examined).[18] But *photographs* of embryos show strong and unquestionable similarities. The embryos of reptiles, birds, and mammals all resemble one another other much more strongly than do the adult forms, exactly as Darwin noted in the *Origin of Species*. Moreover, the similarities are not just superficial. They involve most of the fundamental pathways and structures of embryogenesis. Darwin and Haeckel asked why such different adult forms should all be modifications of what amounts to the same embryological plan—if organisms were specially created, they could just as well each develop directly into the adult forms with no embryological resemblance and no cumbersome remodelings during late embryonic life. Michael Richardson, the specialist who, in an exhaustive critique of Haeckel's work, reexamined all the drawings, observes:

> *On a fundamental level, Haeckel was correct: All vertebrates develop a similar body plan (consisting of notochord, body segments, pharyngeal pouches, and so forth). This shared developmental program reflects shared evolutionary history. It also fits with overwhelming recent evidence that development in different animals is controlled by common genetic mechanisms.*[19]

The cry of "Fake!" from Wells and friends is a completely manufactured scandal.

THE ORIGIN OF INFORMATION IN DNA

The Kansas science education standards state that "the sequence of the nucleotide bases within genes is not dictated by any known chemical or physical law." This assertion is copied from the 1980s creationists who wrote *Of Pandas and People* and later founded the ID movement. Dean Kenyon, for example, included it in his 1984 affidavit in defense of the creationist Louisiana Balanced Treatment Act as it wound its way up to the Supreme Court. It formed the basis of books and articles written by two leading ID proponents: Charles Thaxton, academic editor of the *Pandas* project, and Stephen Meyer.

Their argument: The order of the chemical "letters" in DNA is not dictated by any known physical or chemical law; therefore, the information in DNA cannot be explained by natural processes;

therefore, the information in DNA must have a supernatural cause. Stephen Meyer explains in his chapter for the 1994 anthology *The Creation Hypothesis:*

> *Scientists have attempted to explain how purely natural processes could have given rise to the unlikely and yet functionally specified systems found in biology systems that comprise, among other things, massive amounts of coded genetic information. The origin of such information, whether in the first protocell or at those discrete points in the fossil record that attest to the emergence of structural novelty, remains essentially mysterious on any current naturalistic evolutionary account.*[20]

It would be no exaggeration to say that this argument is at the heart of the ID movement. The only problem is that it is scandalously *wrong*. Competent scientists know how new genetic information arises: a variety of well-understood mutational mechanisms copy and modify the DNA letter sequence that makes up a gene. If the new sequence is advantageous to the organism, natural selection spreads the new gene through the population by way of well-understood processes of population genetics. This shows where new genetic information comes from, and it fully explains, as a bonus, the otherwise puzzling fact that most genes belong to large families and superfamilies of similar composition.

One particularly useful paper was published in *Nature Reviews Genetics* in 2003; written by Manyuan Long of the University of Chicago, it reviews all the mutational processes involved in the origin of new genes and then lists dozens of examples in which research groups have reconstructed the genes' origins. The paper lists 122 references, virtually all of them published in the last ten years. None has ever been mentioned by the ID movement, let alone rebutted. Dr. Long has devoted his whole career to studying the origin of new genes; his online résumé lists some two dozen recent publications on the topic.[21]

The other problem with the argument of the Kansas science standards is the obfuscation "any known chemical or physical law."

It is deviously phrased to have two meanings: it could simply mean that no laws of chemistry or physics specify the order of the chemical "letters" in DNA. In that limited sense, the statement is approximately correct, but pointless. The shape of the Grand Canyon is also not *strictly* specified by any chemical or physical process—so what? The shape of the Grand Canyon is in fact specified by complex but reasonably well-understood interactions of erosion, rock structure, weather patterns, plate tectonics, and the like. But the Kansas science standards statement is meant to imply that no natural explanation exists for genetic information. That is a radical and false claim. Like the Grand Canyon, no simple physical law determines the DNA sequence, but the complex interacting processes that do explain it, described above, are well known. The game being played here is that the radical claim—that there is no natural explanation for DNA—will be taught to students, but when the standard comes up for criticism, the limited but mostly true claim—that we do not know the exact laws of physics or chemistry governing the structure of DNA—will be used to defend the phrase.

IRREDUCIBLE COMPLEXITY AND MICROEVOLUTION/MACROEVOLUTION

The Kansas Science Education Standards state on page 76,

> *Whether microevolution (change within a species) can be extrapolated to explain macroevolutionary changes (such as new complex organs or body plans and new biochemical systems which appear irreducibly complex) is controversial. These kinds of macroevolutionary explanations generally are not based on direct observations and often reflect historical narratives based on inferences from indirect or circumstantial evidence.*

IRREDUCIBLE COMPLEXITY (IC) is a term coined by ID advocate Michael Behe in his 1996 book *Darwin's Black Box*. It is used seriously only by ID advocates and other creationists. ID advocates cannot say truthfully that the Kansas science standards do not mandate ID because the universally favorite ID argument—IC—is explicitly a part of the standards. Furthermore, that argument is just a biochemical version of the now ancient creationist canard, "What good is half a wing?" (Correct answer: "Ask a feathered dinosaur fossil.")

The irreducible complexity argument for intelligent design proceeds as follows: First, the uncontroversial observation is made that many biological systems have multiple required parts. Second, it is asserted that gradual, step-by-step evolution—natural selection acting on natural variations—cannot produce a system with multiple required parts, because, allegedly, any intermediate system lacking all of the required parts would have no function, and therefore would not be preserved by natural selection. The final step in the argument is to conclude that these systems look designed, and to conclude further that if natural selection can't produce these systems, then intelligent design is the only other alternative.

There are severe problems with the second and third steps of these arguments. The most important objection to the second step is that it incorrectly assumes that evolution by natural selection always proceeds in only one way—gradual improvement of an already existing function. But it has been well known, and repeatedly emphasized by evolutionary biologists ever since Darwin, that biological systems commonly *change function*. Consider the flipper of the penguin. It is descended from the flapping wing of a bird, and indeed many sea birds today use their wings for both flying and swimming. In penguins and several other flightless birds, a complete transition to swimming has occurred. As for where bird wings came from, a fantastic series of fossils of feathered dinosaurs has been uncovered in China over the last decade—bird wings are modified versions of the forelimbs of small, bipedal dinosaurs. Moving farther back still, dinosaur forelimbs are modified versions of the front feet of four-legged reptiles. And of course tetrapod feet are ultimately modified versions of the fins of lobe-finned fish. Thus the same fundamental structure was used progressively for swimming, walking, grasping and climbing, flying, and finally back to swimming. In each case we have evidence that the transition could occur smoothly, with intermediate structures performing multiple functions at once. The same point can be made for many other structures. Stephen Jay Gould and Elisabeth Vrba even gave this process a name, *exaptation*, to emphasize the point. The significance for creationist claims about irreducible complexity is clear: the possibility of change of function in evolution vitiates the argument.

Faced with such anatomical counterexamples, Behe tends to assert that his argument is restricted to the molecular level (where fossil evidence is lacking), and that all of the evidence for the evolution of complex systems at the anatomical level should be ignored when examining his argument. This is a specious assertion because irreducible complexity is a scale-free concept. Behe's own favorite example of irreducible complexity—the mousetrap—proves the point: the *size* of a mousetrap doesn't matter, nor does it matter that the wooden base is composed of very complex plant cells. Moreover, creationists and ID advocates before and after Behe have gladly applied the multiple-parts-required argument at all levels of biology.

Nevertheless, Behe has gotten substantial traction from his biochemical examples, because they are wondrously complex and they certainly seem difficult to explain by way of gradual evolution—at least to evolutionists who know little about biochemistry and to biochemists who know little about evolution. But interdisciplinary research reveals that *change of function* is equally important for evolution at the molecular level. This is spectacularly proven by one of Behe's own favorite examples, the adaptive immune system of vertebrates. In the adaptive immune system, invading bacteria and viruses are recognized with protein receptors called immunoglobulins. In order to recognize any possible invader, the immune system produces billions of different receptors by rearranging pieces of receptor genes in a complex process called VDJ recombination. Behe says that VDJ recombination is irreducibly complex and could not have evolved, and in *Darwin's Black Box* he famously asserted, "We can look high or we can look low, in books or in journals, but the result is the same. The scientific literature has no answers to the question of the origin of the immune system."[22] But in fact, back in 1979, seventeen years before *Darwin's Black Box* was published, scientists proposed a hypothesis to explain the evolutionary origin of this DNA rearrangement, suggesting that perhaps a "jumping gene" called a transposon inserted itself into a primitive, nonrearranging receptor gene. When this gene was expressed, the transposon would cut itself out of the gene, but because this process is inexact, the resulting protein product would be variable. Multiple rounds of gene duplication and rearrangement would expand this basic system into the modern version of VDJ recombination.

Behe didn't buy this story, and said so in his book. But unfortunately for him, starting in the mid-1990s immunologists have published discovery after discovery that have confirmed the transposon hypothesis. In 2005, a stunning prediction was fulfilled when a relative of one of the immune system's recombination activating genes, called RAG1, was found as a free-living transposon in tunicates. When Behe was cross-examined during the *Kitzmiller v. Dover* trial in October 2005, he was asked if he still believed that the scientific literature had no answers on the origin of the immune system. He said that he did. Then, in a Perry Mason–like flourish, the plaintiffs' attorney piled fifty-eight peer-reviewed articles and a stack of books about the origin of the immune system on the witness stand in front of Behe. When asked, Behe said that he had not read most of them, but dismissed the pile with a wave of his hand. As the cell biologist Kenneth Miller, who testified for the plaintiffs in the trial, put it later, "Ain't nothing going to convince this guy." Judge Jones, however, was convinced, and noted in his decision that the concept of exaptation—change of function—was fundamental in evolutionary biology, and that if it is given appropriate attention, Behe's arguments fail.[23]

MICROEVOLUTION/MACROEVOLUTION. The claim that microevolution can't be extrapolated to macroevolution is ubiquitous among ID advocates and the creationists who preceded them. It is a central theme in the Kansas science standards and the Ohio critical analysis of evolution lesson plan (originally entitled The Great Macroevolution Debate). But it is nothing more than standard creation science terminology for the creationist claim that various groups of organisms were specially created by God, with specified limits on how far they could change over time. Creationists say that macroevolution—which they define as transformations among biblical kinds—just can't happen. Skeptics search in vain for any objective definition of *kind*—in practice, creationists make the definition strategically elastic. Any evidence for evolution so overwhelming that even creationists cannot deny it is labeled *microevolution within a kind*. Macroevolution of new kinds, conveniently, is never attained.

This smoke-and-mirror tactic is at the heart of the creationist enterprise. It persists unchanged among creation science, intelligent design, and critical analysis, and is uniformly employed to protect a single thing: the biblical doctrine of special creation of the Genesis kinds (and particularly of the human kind). At the Kansas Board of Education hearings on evolution, witness after witness—all allegedly respectable scientists testifying in support of critical analysis of evolution—stammered out denial of common ancestry when asked the question. As the microevolution/macroevolution distinction is a central feature of all three antievolution tactics, its origins and problems will be explored in more depth.

Special Creation in Disguise

THE CREATIONIST ROOTS OF MICROEVOLUTION VERSUS MACROEVOLUTION

The creationist Frank Lewis Marsh, writing in the *Creation Research Society Quarterly* in 1969, carefully defined the two categories, "microevolution and *megaevolution* (also called *macroevolution*)" as follows:

> Microevolution *is the term applied to the demonstrable production of new varieties or breeds within any basic type.*
>
> Megaevolution, *on the other hand, is the term applied to the doctrine which holds that, if given enough time, basic types can eventually, through natural processes, produce new basic types.*

Microevolution, says Marsh, is "demonstrable variation." He finds it unfortunate that microevolution is called evolution at all, "because the special creationist who accepts all demonstrable variation does not accept the theory of evolution." Marsh lists fifteen breeds of cow and six kinds of corn and calls them all products of "only microevolution." To emphasize the point he even says that microevolution can produce new species. But, says Marsh, it "can never accomplish megaevolution."

Marsh used the micro/macro distinction in defense of the fundamentalist belief that the kinds of animals described in the Book of Genesis were specially created by divine action. After reviewing the days of creation in the Genesis story, Marsh focuses on the Bible's repeated statement that organisms reproduce after their kind, writing,

> *The record states, in Genesis 1:11, 24, that the command to the earth was to produce these organisms* after their kinds. *In other words these forms were first conceived in the mind of God, and then at the fiat to the earth to bring forth, the Spirit of God (Genesis 1:2) working with the dust of the earth (Genesis 2:7, 19) produced from it the living forms patterned according to the plan.*

Marsh recognizes that variation from type occurs, and that it can and does lead to change of type, but that is not enough to overcome the biblical account of creation: "Variation does occur abundantly *within* kinds, but no coercive, compulsive evidence can be produced to show the production of even one new basic kind." He dismisses common examples of evolution, such as the remarkable diversity of finches Darwin found in the Galápagos. These are, in Marsh's view, mere variations within the kind: "Darwin failed to recognize the tremendously important fact that the tortoises were still tortoises and the finches still finches."[24]

Similar sentiments are ubiquitous in creationist literature. In an informal search of *Creation Research Society Quarterly*, we were able to find twenty examples between 1970 and 1976 of creationists belaboring the microevolution/macroevolution distinction. For example, a 1970 article states,

> *A fruit fly or lamp shell after mutation is still a fruit fly or a lamp shell. Does "microevolution" prove "macroevolution"? i.e., does the production of a new variety of fruit fly or lamp shell explain the creation of the fruit fly or lamp shell itself?*

The universal creationist reaction to the peppered moth, which evolved its camouflage from white to black when the background trees were darkened by industrial soot, is to state that this is just microevolution. This is very often accompanied with an accusation of evolutionist dishonesty, for example,

> *Actually [the peppered moth] is not an example of evolution at all. The moths are just the same as they were before the industrial revolution and are not in a process of becoming anything else. Thus it may be charged that referring to such phenomena as any kind of evolution (as micro-evolution) amounts to a brainwashing technique to make real evolution seem less objectionable to those who have scruples against the concept of evolution.*

The actual point of the peppered moth example—that it illustrates how camouflage, a common adaptation that appears designed, can evolve through a simple natural process—is always completely ignored.[25]

WHAT'S WRONG WITH THE CREATIONIST DEFINITION OF MACROEVOLUTION

To a biologist, the "it's just microevolution" argument is painfully obtuse. In normal science, microevolution refers to evolutionary processes *within* gene pools, such as the origin and spread of individual gene variants. Macroevolution refers to evolutionary processes that work across *separated* gene pools. Speciation, a process that can be observed in nature, and that creationists accept, is the boundary between microevolution and macroevolution, because speciation occurs when one gene pool permanently splits into two separate gene pools. A speciation event is a case of macroevolution. So are other events that apply to whole gene pools, such as extinction.

For biologists, then, the microevolution/macroevolution distinction is a matter of *scale of analysis,* and not some ill-defined level of evolutionary newness. Studies that examine evolution at a coarse scale of analysis are also macroevolutionary studies, because they are typically looking at multiple species—separate branches on the evolutionary tree. Evolution within a single twig on the tree, by contrast, is microevolution.

It is true that scientists themselves contribute to confusion over this issue. This typically occurs because *macroevolution* is such a broad term that it can be applied to a wide range of proposed processes, ranging from uncontroversial (extinction, speciation, adaptive radiation, ecological drift), to controversial (punctuated equilibria, species selection), to discredited (orthogenesis, saltation). In a perfect world, scientists would refer to these specific processes rather than the very general micro/macro distinction, but as long as the terms are being used, it behooves us to understand what they mean within the scientific community. Evolutionary biologists on both sides of famously contentious debates seem to agree that the definition of macroevolution boils down to "evolution above the species level."[26]

WHAT'S WRONG WITH THE IDEA OF CREATED KINDS

Once we get past the definitional distraction, we are left with the creationist assertion that there are groups of organisms called kinds, and that as much evolution as you like can occur within these kinds, but that evolution can never, ever, cross the border of the kind. Unfortunately, *the creationists have offered no rigorous criteria for determining the limits of a kind.* The only constant rule seems to be that humans are their own kind.

Indeed, the creationists have been promising rigorous criteria for decades, but they continue to insist that more research is needed. Even the cutting-edge creationists who call themselves baraminologists—*baramin* is a term for created kinds introduced by Frank Marsh, based on the Hebrew words for "created" and "kind": *bara* and *min*—admit that they have a long way to go. When they actually do empirical work, they tend to posit massive baramins. For example, they have determined baraminologically that the entire horse fossil series, from the twelve-inch-high fossil *Hyracotherium* to the modern horse, is all one created kind and therefore the progeny of a unique common ancestor. A baraminological analysis of the group of plant species related to sunflowers found that this created kind contains 5730 species, and the creationist researchers conclude further that "we still cannot rule out the possibility that all 20,000 species of the Asteraceae represent a single holobaramin [i.e., created kind sharing common ancestry]." Asteraceae is the sunflower family, and it is one of the two largest families of flowering plants, competing with the orchids. The family contains annuals, perennials, stem succulents, vines, shrubs, and trees, including everything from lettuce to the dwarf fluffweed of the southern California desert (whose adult plant is one inch high and one inch wide) to twenty-five-foot trees.

As the historian of creationism Ronald Numbers notes, the strictest creationists of all, the young-Earth creationists, now tell us that macroevolution is impossible—except, ironically, for the massive amount of extremely rapid evolution that they propose as having occurred just after Noah's Flood, producing tiny weeds and trees from a common ancestor, and producing modern horses from foot-high precursors.[27]

In the creationist concept of created kind—and the creationist demand to "Show me macroevolution"—we have a classic example of the movable-goalposts strategy for winning. Any amount of evolution that can be demonstrated to the creationists' satisfaction is effectively by definition microevolution within a kind. No matter how extensive the documented change is, the macroevolution goalposts are always out of reach. The inviolable biblical kind is protected with strategic vagueness.

SPECIAL CREATION IN INTELLIGENT DESIGN

The ID movement adopted the microevolution/macroevolution distinction unchanged from its young-Earth creationist inventors, except for hiding the biblical source and the motivation of their insistence upon it. Phillip Johnson, godfather of the ID movement, wrote in *Darwin on Trial*,

> *Everyone agrees that microevolution occurs, including creationists. Even creation-scientists concur, not because they "have tightened their act," but because their doctrine has always been that God created basic kinds, or types, which subsequently diversified. . . . The point in dispute is not whether microevolution happens, but whether it tells us anything important about the processes responsible for creating birds, insects, and trees in the first place.*[28]

Unfortunately for Johnson, even the young-Earth creationists are forced to admit that trees *can* be produced by microevolution as they define it, as we saw in the baraminological treatment of sunflowers reported above.

The term *macroevolution* occurs fifty times in the ID textbook *Of Pandas and People,* and in some forty of those instances (excluding headings, index entries, and so on) it is contrasted with verifiable microevolution—or otherwise disparaged. For example,

> *What breeders accomplish is diversification within a given type, which occurs in microevolution. What is needed is the origin of new types, or macroevolution.*

Predictably, the peppered moths get their microevolution label:

> *Before taking a closer look at the genetic world, it will help us to recognize that there are two levels or categories of evolution,* ***microevolution*** *(small-scale evolution) and* ***macroevolution*** *(large-scale evolution). The shift in populations of moths from light coloring to dark illustrates microevolution.*

Even apart from the now-famous creationist drafts of *Pandas*, it is easy to show that the assertions in *Pandas* are simply the fundamentalist Christian view of special creation, relabeled for public school use. Nancy Pearcey wrote the long overview chapter in *Pandas*. She is currently a fellow at the Discovery Institute's ID program, but in the 1980s she was a contributing editor for the young-Earth creationist *Bible-Science Newsletter*. She republished most of the *Pandas* overview chapter in three articles in the *Bible-Science Newsletter*, with minor sentence rearrangements, and with the systematic omission of creationist terminology from *Pandas*. Although special creation is implicit in *Pandas*, the *Bible-Science Newsletter* version of the chapter contains some additional text so that one doesn't have to read between the lines:

> *Darwin believed change is unlimited, that species are infinitely plastic. He thought a species could vary indefinitely and in any direction. Creationists believe change is limited by a basic organic unit, the created "kind." Within the boundary of that fundamental unit, variation can be profuse. But it never leads to the creation of a new basic type.*
>
> *The fact that the organic world fits a hierarchical pattern constitutes evidence against the evolutionary view and in favor of a theory of separate creations.*

Furthermore, each *Bible-Science Newsletter* article is followed by a Bible-study article that discusses the various Bible passages relevant to the preceding scientific article. For example, Pearcey's BSN article "Of Fins and Fingers" is followed by an article entitled "Bible Study: Which Is More Scientific: 'Kinds' or 'Species'?" Here, Pearcey lays out the theological interpretation on which she is basing the microevolution/macroevolution distinction:

> *Begin by reading Genesis 1:11–25 to make sure that you see both the creation according to kinds for each creature, and the reproduction according to kinds. How many times does the word "kind" or "kinds" ap-*

pear in this section of Scripture? Why is this concept repeated so often in this section?

Would God have known that man would eventually try to explain life in evolutionary terms?[29]

Frank Lewis Marsh himself could not have put it better for a biologically challenged readership.

SPECIAL CREATION IN CRITICAL ANALYSIS OF EVOLUTION

We found thirty-four pages on the Discovery Institute's Web site containing both *microevolution* and *macroevolution*. For example, a FAQ sheet called "Summary: The Scientific Controversy over Whether Microevolution Can Account for Macroevolution" states,

The scientific controversy over whether processes observable within existing species and gene pools (microevolution) can account for large-scale changes over geological time (macroevolution) continues to this day.[30]

This FAQ is a stew of misinformation and quotes taken out of context. For example, it quotes an article on evolutionary developmental biology (evo-devo) discussing the relative importance of micromutations and macromutations, and uses this to support the claim of a controversy about macroevolution. However, macromutations, like any mutations, occur within a gene pool and would therefore be *micro*evolutionary phenomena analyzable within population genetics. Indeed, the whole point of evo-devo has been to bring developmental biology into the evolutionary synthesis with population genetics. The other two quotations simply refer to the distinction between population genetics within lineages and the dynamics of distinct lineages.

Since 2002, the Discovery Institute has been pushing hard for states and local school districts to require the teaching of false controversies like this; their 2005 FAQ lobbies as follows:

> *Since the controversy over microevolution and macroevolution is at the heart of Darwin's theory, and since evolutionary theory is so influential in modern biology, it is a disservice to students for biology curricula to ignore the controversy entirely. Furthermore, since the scientific evidence needed to settle the controversy is still lacking, it is inaccurate to give students the impression that the controversy has been resolved and that all scientists have reached a consensus on the issue.*

What would really be a great disservice would be for the state to require teachers to teach the falsehood that there is any scientific doubt about common ancestry, which is the true target of the creationist microevolution/macroevolution argument. Unfortunately, it is clear that this is exactly what the Kansas Board of Education is doing. Steve Abrams, one of the creationist leaders on the Kansas Board of Education, said it clearly in an op-ed ironically entitled "Science Standards Aren't about Religion."

> *That is one of the reasons that we tried to further define evolution [in the Kansas standards]. We want to differentiate between the genetic capacity in each species genome that permits it to change with the environment as being different from changing to some other creature. In our science curriculum standards, we called this microevolution and macroevolution—changes within kinds and changing from one kind to another.*[31]

Conclusion

Although supporters of strong science education should not expect that the courts will always ride to the rescue, there are reasons to be hopeful in the case of critical analysis of evolution. First, although critical analysis as the creationists see it consists not of critical analysis in the ordinary sense but of a haphazard collection of objections to evolution, it emerges on closer inspection that *every single* critical analysis argument can be traced directly to established, long-discredited creation science and intelligent design claims. This is important because courts have repeatedly emphasized the relevance of the historical lineage of governmental policies. A court that is made aware of this history should and may well conclude that critical analysis of evolution is substantially the same as creation science and intelligent design and see it as one more in an unending series of attempts to privilege the same sectarian view—creationism—with a new and deceptive label.

Second, derived as they are from creationist literature, critical analysis arguments share all the problems of their creationist ancestors. The objections to evolution are not serious scientific arguments; they are superficially investigated and poorly reasoned talking points. Like all creationist arguments, they are aimed at uninformed audiences: they sound good in op-eds, media sound bites, and sermons, but they disintegrate upon detailed examination in court or in serious and extended public discussion.

3: Theology, Religion, and Intelligent Design

MARTINEZ HEWLETT AND TED PETERS

Although many interpret the controversy over the teaching of evolutionary biology in public and private schools as a battle between science and religion, we believe this is a mistaken perception. At the deepest level, no conflict between genuine science and healthy religion exists. All parties to the controversy have a positive view of science, even outspoken religious debaters. Virtually all religious voices are raised on behalf of good science, even though what constitutes the best science is a matter of dispute. It is the view of the two authors of this chapter—one a Lutheran theologian and the other a Roman Catholic biologist—that the standard model of evolutionary biology constitutes the best science. Further, we contend that evolutionary biology—even when understood as the Darwinian or neo-Darwinian model—should be taught in science classes in both public and private schools. We recommend against giving equal time to intelligent design (ID) and to scientific creationism. We deem both ID and creationism to be unsatisfactory models for scientific research; further, they are not even science at all, as we understand the discipline. It is incumbent on us as a society to offer our young people only the best science, and the best biological science in our era proceeds from the recognized theory of evolution complete with random variation and natural selection.

Why do we take this position? Because we believe that science at its best and religious commitment at its best honor truthfulness. Further, we believe that religious communities and scientific research along with science education must cooperate for the betterment of our society.

We take this position based upon theological commitments. We do not feel compelled to protect secular science from alleged religious contamination, as if religion might be something dirty that would smudge scientific purity. Rather, we have a high opinion of religious commitment, especially intellectually healthy religious thinking. Our argument follows a theological path. As Christians, we believe our faith demands the best science; the best biology follows in the tradition of Darwinism and neo-Darwinism and subsequent emendations to Darwinian theory. Faith seeks understanding (*fides quaerens intellectum*) said Saint Anselm eight centuries ago, and right now the best understanding of living things requires the theory of evolution. In short, we support evolutionary biology as an expression of our religious faith, not in contradiction to it.

We are not suggesting that standard evolutionary biology become a dogma of faith, to be sure. Science should remain science, even when pursued by people of faith. Like all science, the theory of evolution is subject to critical review and revision over time. Based on our theological support for the best science, we embrace this ongoing critical process and fully expect to see expansion and deepening of our understanding of biological life in the generations to come.

One of the disturbing aspects of the current controversy is the misleading impression that if one is religious, one must be antievolution, even antiscience. It pours fuel on the fire of an already burning controversy. We believe that intellectually, no conflict between science and religion exists. It's a fake conflict. Some with vested interests want to keep the conflict going, however, so they keep drawing pictures of warriors in skirmishes. According to one picture, the warriors for science appear to be dispassionate champions of uncompromising civility and truth, while warriors for religion appear to be atavistic throwbacks to premodern ignorance. According to another picture, establishment scientists appear to be dogmatic defenders of atheistic materialism, while warriors for religion are champions of free speech and high-minded moral values. These are distorted pictures. Yes, these pictures may in fact describe certain colorful individuals engaged in the culture war. But they do not reflect an honest assessment of the intellectual issues at stake between a healthy commitment to faith and a healthy commitment to scientific knowledge.

As we have indicated, the two of us embrace healthy science as an expression of our religious faith. Hewlett is a Roman Catholic who has been engaged for decades in researching and teaching molecular biology with a special focus on virology. Peters is a Lutheran pastor and professor of theology with a special interest in the dialogue between faith and science. We place ourselves roughly in the camp of *theistic evolution*—that is, we believe Christian faith and evolutionary theory are compatible. More important, the curiosity that leads to the pursuit of science is an expression of our Christian faith, not a hurdle to be jumped or a blockade to be avoided. The last thing we would want to see in our public or parochial schools would be a *sacrificum intellectus,* a sacrifice of the intellect, in order to protect religion.

It is also important to point out that non-Christian religious traditions are less likely to take a stand against evolutionary biology in general and the Darwinian model in particular. Anti-Darwinism does not attract non-Christian religious leaders, at least for the most part. Jewish thinkers made their peace with evolutionary theory a century ago. Although some Turkish Muslims feel a kinship with Christian anti-Darwinism, overall Islam has no beef with evolutionary biology. New Age spirituality has wholeheartedly embraced the evolutionary vision and incorporated it into a futuristic vision where matter and spirit become united. From the perspective of other parts of the world, the skirmish over intelligent design looks like North American parochialism.

In what follows we will review the position advanced by intelligent design. We will treat ID as a theological proposal along with its claim to be a scientific proposal. We will say something about ID assumptions regarding the relationship of the Intelligent Designer to the natural world in theological terms. Then we will look briefly at the stake Jewish and Muslim theologians have in the evolution controversy.

The Theological Concept of Design before ID Came Along

The key idea in the controversy as the intelligent design advocates formulate it is this: our observation of design in nature suggests the existence of an Intelligent Designer that transcends nature. Now, the contemporary discussion over detectable design in nature has a lengthy history that predates today's intelligent design school of thought. Our philosophical and theological ancestors, such as Aristotle, Thomas Aquinas, and William Paley, all found design if not purpose in nature, and this led them to postulate if not attempt to prove the existence of God. God is the all-intelligent One Who created a rationally operating world that reflects the divine mind. By looking at design in nature, we catch a partial glimpse of the otherwise invisible author of nature.

We draw something different from this ancestry than the intelligent design school does. From Thomas Aquinas, we draw on the distinction between primary and secondary causality. God is the primary cause. God created the world. God continues moment by moment to sustain the existence of the world and to prevent it from dropping into nonbeing. What happens within the world, however, happens according to secondary causes. God created the world so that it could self-organize. What scientists study are secondary causes, not God as the primary cause.

What the intelligent design school draws upon is the link between design within the realm of secondary causation and primary causation. William Paley (1743–1805) sets the precedent for the contemporary discussion in his important book *Natural Theology: or, Evidences for the Existence and Attributes of the Deity, Collected from the Appearances of Nature.*[1] Charles Darwin read it. No doubt that when Darwin later developed his theory of evolution he remembered Paley. What he and the rest of us remember is Paley's argument from design, most tellingly illustrated by his analogy of the watch and watchmaker.

> *In crossing a heath, suppose I pitched my foot against a* stone *and were asked how the stone came to be there, I might possibly answer that for*

> *anything I knew to the contrary it had lain there forever; nor would it, perhaps, be very easy to show the absurdity of this answer. But suppose I had found a* watch *upon the ground, and it should be inquired how the watch happened to be in that place, I should hardly think of the answer which I had before given, that for anything I knew the watch might have always been there. Yet why should not this answer serve for the watch as well as for the stone? Why is it not as admissible in the second case as in the first? For this reason, and for no other, namely, that when we come to inspect the watch, we perceive—what we could not discover in the stone—that its several parts are framed and put together for a purpose . . . the inference we think is inevitable, that the watch must have had a maker—that there must have existed, at some time and at some place or other, an artificer or artificers who formed it for the purpose which we find it actually to answer, who comprehended its construction and designed its use.*

Paley's watchmaker is both designer and craftsman in this case —the one who designed and built the watch. In like manner, Paley continues, such features of the biosphere as the human eye must also be objects of purposeful design. And he concludes that the designer is none other than God. The kind of God affirmed by Paley is the God of theism, one who creates the world at the beginning and then intervenes within the subsequent world processes to move creation forward.

What follows from this analogy is an argument that begins with marveling at the wonders of complexity in nature. The human eye, for example, is a complex system designed for seeing. The various parts of the eye cannot see by themselves. Only through a complex pattern of interaction of the various parts is the eye as a whole ready to fulfill its purpose, namely, to provide sight for the organism. Isn't God marvelous for having designed the eye?

Charles Darwin was not satisfied with Paley's arguments, or with the role of an interventionist God in biological evolution. Darwin himself tried to replace Paley's arguments for a transcendent designer with a strictly immanent or noninterventionist explanation: the principles of variation in inheritance combined with natural selection in the creation of new species sufficiently explains the evolution of new life forms. The replacement of interventionist divine action with a strictly natural explanation appeared, at the end of the nineteenth century, to pit science against religion. Darwin himself did not believe that his theory refuted the existence of God; he merely limited his science to an admittedly this-worldly explanation. The question of God's existence remained untouched by Darwin. Thomas Huxley, however, used Darwin's model to justify agnosticism and even atheism. Many of today's Darwinists follow Huxley, not Darwin, in this matter.

The British champion of secular science and atheistic materialism Richard Dawkins confronts Paley's theological position in his book *The Blind Watchmaker.*[2] Dawkins admits that design is observable in the natural world, especially in the case of such elaborate structures as the vertebrate eye. However, Dawkins holds that such design evolved by gradual steps over deep time, and that there is no direction to this process, no purpose. All design is exhaustively explainable by natural causes, by random variation in genetic mutation and natural selection. No intelligent designer is necessary. Evolutionary biology expunges any role for an interventionist God in the development of new species. Dawkins embodies everything that ID fears: Darwinian theory, materialism, and atheism, all conspiring to have a designed world minus any designer. The question for our era is this: can we retrieve Paley's argument from design to combat a secular materialism based upon Darwinian theory? ID says yes. We say no.

Intelligent Design Is Born

The intelligent design or ID school of thought is the new toddler in the schoolyard of the evolutionary controversy. Its ancestry goes back to Greek and medieval ways for knowing God and to William Paley's use of design to prove God's existence. But this is only one side of the family lineage. The other side includes other anti-Darwinists, such as some outspoken fundamentalists of the 1920s and the scientific creationists of the 1970s and 1980s. Although the

ID brotherhood is clearly distinguishable from its ancestors on both sides, it retrieves previously used arguments from design, polishes them up, and repackages them to give to contemporary discussion. This is an observation, not a criticism; many of today's best ideas were introduced by ancient geniuses in our history.

ID spokespersons dissociate themselves from the side of their family history that leads back toward fundamentalism and creationism. Even if it is related, the position ID espouses is quite different. Whereas fundamentalism established its position on the authority of the Bible, ID claims to base its contentions on scientific observation and theory. Whereas scientific creationism has contended that God created all the kinds of life at the beginning, thereby fixing the species and denying speciation over time, ID, in contrast, can affirm the history of speciation and simply appeal to transcendent intervention at crucial stages within evolutionary history. Theologically, fundamentalists and scientific creationists deal with creation, whereas ID deals with providence within creation. Further, creationists call the original creator "God," while ID avoids theistic terminology and refers only to an Intelligent Designer.

Despite these sibling rivalries, both creationism in its various forms and intelligent design unite as a family against the dominant Darwinian model taught in the public schools. In the heat of controversy, especially in the courts, however, ID disowns its family relationship with fundamentalism and even creationism. This is because of the relationship of ID to the First Amendment of the U.S. Constitution. Do proposals to teach ID violate the proscription against government establishment of a sectarian religious position? By appealing strictly to scientific argument and by refusing to connect the Intelligent Designer with what others call God, ID advocates believe such proposals can escape condemnation as violations of the First Amendment.

The constitutionality of teaching intelligent design in the public schools is not the matter we wish to analyze here (see chapter 4 for further discussion on that point). Rather, we would like to look at ID theologically and in light of scientific criticisms. We suspect that the commitment ID makes to anti-Darwinism is unnecessary, and we do not recommend that ID science be placed on a par with standard evolutionary science in the classroom. We sympathize with ID insofar as it is waging a battle in the culture war against belligerent atheism; but we believe intellectually that affirmation of natural explanations for evolution are compatible with a healthy faith in God as Creator and Redeemer.

Our plan now is to give ear to three voices within the ID choir, Phillip Johnson, Michael Behe, and William Dembski. Each is worth listening to. Even so, we have our criticisms. We will offer criticisms following a brief exposition.

Phillip Johnson's Case Against Darwin

In 1991, Phillip Johnson published *Darwin on Trial,* a polemic against both the science and the court proceedings surrounding the teaching of evolution in U.S. schools. Phillip Johnson is a professor emeritus of law at the University of California, Berkeley. His legal specialty is criminal law, but more recently his attention has turned to evolution and naturalism as a philosophical system.

Important to note is that Johnson is not trained as either a scientist or a theologian, nor has he ever practiced either discipline. His analysis of evolution is therefore based upon his own reading of the lay literature to which he has access and the interpretation of the scientific literature by popularizers. As a result, neither this book nor his subsequent ones provide a satisfactory scientific critique of biological evolution. Nor does it break new ground theologically. Nonetheless, its publication led to a large following, and he has had an active career on the lecture circuit as a result. What Johnson offers to the ID movement is an anti-Darwinism that is aimed at jumping the legal hurdles of the First Amendment.

Johnson's arguments against the scientific claims of evolution focus on the gaps in the fossil record, the lack of what he considers credible evidence supporting speciation, and difficulties of classification raised by the newer molecular analyses of existing species. For example, he challenges what he refers to as the fact of evolution by calling into question any support for macroevolution (evolution from one species to a new species or speciation) as a central tenet of Darwinism. In response to the microevolutionary (evolutionary change within a species) examples cited by Darwinists, Johnson retorts:

> *But what sort of proof is this? If our philosophy demands that small changes add up to big ones, then the scientific evidence is irrelevant . . . What Darwinists need to supply is not an arbitrary philosophical principle, but a scientific theory of how macroevolution can occur.*[3]

Note that this is legal rhetoric. It does not constitute an alternative scientific model to explain the origin of species. Johnson's self-appointed task is not to come up with valid scientific objections or alternative explanations for biological evolution. Rather, he is examining the evidence in a legal sense and passing a judgment based on that examination.

Such rhetorical arguments that focus on the weakness of the dominant scientific paradigm are reminiscent of what one finds among the scientific creationists, especially the work of Henry Morris and Duane Gish.[4] One could argue from a reading of *Darwin on Trial* that, in spite of his disassociation from creationism, he still belongs in the family.

Michael Behe's Irreducible Complexity

Darwin's Black Box: The Biochemical Challenge to Evolution is Michael Behe's principal contribution to the public controversy. Michael Behe is a professor of biochemistry at Lehigh University. He had an active research program on the structure of DNA and histones, proteins that interact with DNA. However, his career has taken a dramatic turn from the laboratory toward the public with the publication of this book.

Behe recognizes what all scientists are aware of, namely that the nineteenth-century Darwin did not have the benefit of twentieth-century knowledge about how a cell or DNA works. As the science of genetics developed, it was combined with Darwin's theory. Genetics contributed an explanation for random variation in inheritance, and Darwin contributed the idea of natural selection to what we now know as the neo-Darwinian synthesis. Today's biologists build on and expand this synthesis.

Darwin, says Behe, did not know about the internal working of cells, the details of which have been elaborated during the twentieth century. Therefore, looking at evolution from the aspect of whole organisms is different from examining the complexities of intracellular pathways. Where does this observation lead?

Behe's central thesis is that there exist structures that are irreducibly complex. By *irreducible complexity* he means that the structure will not function if one part of it is removed:

> By irreducibly complex *I mean a single system composed of several well-matched, interacting parts that contribute to the basic function, wherein the removal of any one of the parts causes the system to effectively cease functioning.*[5]

The implication of irreducible complexity is that complex systems—such as Paley's watch or the eye—could not have developed gradually, step by step, over deep time. Such structures could not have arisen by standard Darwinian gradual processes.

Instead of Paley's watch, Behe uses a mousetrap analogy as his model for an irreducibly complex structure. How could evolution produce such a structure by gradual, single step changes? What would the selection pressure be for the function of the spring in the absence of the catch? He asks these same rhetorical questions about six biochemical systems and cellular structures: the eukaryotic cellular cilium, the bacterial flagellum, the blood-clotting cascade, the intracellular trafficking system, the immune system, and pathways of intermediary metabolism. In each case, he describes the system in detail and argues that it could not have arisen by standard gradualist mechanisms in the Darwinian model, again using his argument that such systems are irreducibly complex. He then concludes that there must have been an Intelligent Designer who arranged each of these systems with a specific purpose in mind.

William Dembski's Mathematical Formulation

Johnson is a lawyer, and Behe a biologist. The next voice we will listen to comes from William Dembski, a person with earned doctorates in mathematics from the University of Chicago and in philosophy from the University of Illinois at Chicago, plus a master of divinity from Princeton Theological Seminary. He is currently a research professor of philosophy at Southwestern Theological Seminary. His first book, *The Design Inference,* opened a different approach to the intelligent design discussion. He followed this with a more detailed exposition in *No Free Lunch.* Other books have cascaded into the market since.

As with Behe, Dembski's strongest suit is complexity, although he plays it from a mathematical and philosophical hand. In *No Free Lunch,* Dembski sets out a method for determining if a given event or object or structure is designed. To be designed, something must be complex. What he observes as designed he dubs "specified complexity." This is not simply a restatement of Behe's irreducible complexity. Dembski produces an algorithm that, he argues, allows one to calculate whether or not specified complexity exists in a particular case. This is how the explanatory filter works. The detection of specified complexity requires that the filter answer three questions: Is the event or object contingent? Is it complex? Is it specified?[6]

Contingent events cannot be explained by the operation of physical necessity, such as the properties of water leading to the formation of ice crystals. Complexity means that the event or object cannot be explained by chance. In this regard, Dembski introduces a universal probability bound of 1 in 10^{150} as the limit below which it is improbable that chance was the source of the event or object. This number is a product of the estimated total number of elementary particles in the universe (10^{80}) times the age of the universe in seconds (10^{25}) times the number of events that can occur per second (10^{45}, or the Planck time). Finally, an event or object is said to be specified if it matches a pattern that is "detachable" from the event or object itself. This condition is one that Dembski spends a good deal of time discussing in *No Free Lunch* and one that has engendered much criticism. Dembski states this criterion as follows:

> *Detachability can be understood as asking the following question: Given an event whose design is in question and a pattern describing it, would we be able to explicitly identify or exhibit that pattern if we had no knowledge which event occurred?*[7]

This is a critical part to the argument, since answering no to this question allows the explanatory filter to relegate the event or object to the category of chance. This has become one of the focal points for critiques of Dembski's explanatory filter. In any case, Dembski contends that events that are contingent, complex, and specific are therefore designed. Such design requires a designer, he says.

What happens next? What Behe and Dembski have done is suggest that the Darwinian model that relies exclusively upon random variation and natural selection to explain the long, slow, step-by-step path of evolution is inadequate. A strictly natural explanation is inadequate to explain the appearance in nature of irreducible complexity or specified complexity. What is needed is appeal to a transcendent designer who intervenes in otherwise natural processes. What is needed is appeal to saltations or interventions into the path of development to account for sudden leaps to higher levels of complexity. Without naming the intelligent designer "God," this school of thought nevertheless finds an opening in scientific theory for appeal to transcendent reality.

Because the appeal to transcendent reality is inherent in explicating what we observe in nature, we do not need to rely upon the special revelation of theology, say the advocates of ID. We do not need to rely upon previously developed religious points of view. We need only rely on what science itself provides. For this reason, ID advocates believe that their alternative scientific model breaks beyond the limits of the Darwinian model. What this indicates is the intelligent design advocates work with a high regard for science, so high that they rest their case for an intelligent designer on science rather than divine revelation.

Our Critique of Intelligent Design

Criticisms of intelligent design come from many directions. One direction is the *sneaky religion argument*. Some argue that ID is only fundamentalism or creationism in disguise, and that ID wants to slip God and miracles under the door of our high school science classrooms. If we want to keep our schools secular, then we need to unmask the deception and relegate ID to its proper sphere, religion. We need to redefine ID so it appears to violate the First Amendment, then we can use the force of law to outlaw its influence. This critique says ID is really religion and not science; therefore, it must be excluded from science classrooms. At least for some, the assumption here is that a war is being fought between religion and science, and excluding ID is a way of winning the battle for science.

Another direction is the *bad religion argument*. Some within the more liberal branches of Christianity believe intelligent design merely represents the present generation of fundamentalists. What is wrong with fundamentalists is that they interpret the Bible woodenly, literally. They mistakenly treat the Bible as a scientific textbook. What ID fails to see is that science and religion speak two separate languages; these two operate in two separate spheres. They do not conflict, because they are independent. ID advocates don't accept this, so they represent a throwback to premodern Christianity before the peaceful separation between science and religion took place. The assumption here is that a war is being fought between conservative and liberal Christianity; and a defeat for ID is a victory for the liberals.

Both of these critiques assume that we are engaged in a culture war, and both want to see ID defeated. The writers of this chapter have no reason for taking sides in any culture war. Neither of these two views represent where we are coming from. We are quite critical of the intelligent design movement; but our concern is with what constitutes healthy science and healthy faith.

In our pursuit of healthy science, we observe how Johnson, in *Darwin on Trial,* bases his criticism of evolutionary theory on what he calls an examination of the science upon which it is based. As he states:[8]

> *My purpose is to examine the scientific evidence on its own terms, being careful to distinguish the evidence itself from any religious or philosophical bias that might distort our interpretation of that evidence.*

The word *evidence* has two meanings, one legal and the other scientific. To conflate these two meanings is fallacious.

In this quoted statement, Johnson intends the word *evidence* to have the same kind of meaning that is given in a legal setting. That is, he looks at the science as a set of logical statements that he can, as an academic lawyer, reason out in terms of their value. The first problem arises when it becomes apparent that Johnson does not fully understand the character of scientific evidence, in terms of the methodology and the data. For instance, Johnson uses the word *random* in a variety of ways in his discussion of mutational changes. One of his scientific errors is to assume that random change means that virtually any kind of change is possible. He states by analogy:

> *A random change in the program governing my word processor would easily transform this chapter into unintelligible gibberish, but it would not translate the chapter into a foreign language, or produce a coherent chapter about something else.*[9]

Now, if Johnson examined all of the variants that his word processor produced, he could conceivably find one that was in perfect Spanish. This process is called selection, and it is this feature of evolution that Johnson ignores in his analysis of mutations.

Second, we believe Johnson confuses science as a method with *scientism* as an ideology.[10] By *scientism* we mean the metaphysical use of scientific information to create a philosophical worldview that relies upon materialism or naturalism or even atheism. Johnson is on target when he criticizes certain scientists who speak or write about their ideological interpretation of Darwinian evolution, when they draw philosophical inferences that are not consequences of the science *qua* science. We distinguish between science as research and scientism in the form taken by Darwinian ideologies (such as social Darwinism, eugenics, materialism, and more recently, sociobiology). These latter four are philosophies that use Darwin for support but that do not constitute the kind of science that laboratory researchers rely on or the kind of science taught in our school classrooms. However, Johnson does not explain why rejection of ideology should be grounds for rejecting the entire body of evidence presented by the scientific method. Is Johnson throwing the baby out with the bath water?

Third, we do not accept Johnson's analysis of the question of common descent. Do human beings share a common ancestry with other primates? On this issue, we must be critical of Johnson. He rejects compelling evidence gained from genomic analysis that shows the genealogical relationships of living systems at the level of protein structure. As support for this rejection, he mentions the apparent conflicts between the sequence analysis data and the fossil record. As an example, he cites comparisons of the protein cytochrome c found in bacteria with the version found in higher organisms (a number of different plants and animals). He notes correctly that the sequence divergence ranges more than 60 percent between the bacterial cell and the higher organisms. He then concludes that this does not represent genealogical linkage, since no intermediates have been located. Johnson states that this invalidates the molecular data as any support for the Darwinian model. This is the same argument that he and others use against the fossil record.

Here's the problem. Johnson's line of reasoning does not recognize the extraordinary correlation between the molecular data and the model itself, especially given the evolutionary distance between prokaryotes and higher eukaryotes. Terry Gray, formerly a professor of chemistry at Calvin College and now working for Colorado State University, has criticized this position with respect to the specific case of a nonactive gene for vitamin C synthesis in primates. It has been discovered that all mammals except guinea pigs and primates have the gene for making vitamin C. However, analysis of the genome of those two organisms shows the presence of an inactive pseudogene for vitamin C synthesis. Gray says:

> *Now we could argue that in God's inscrutable purpose he placed that vitamin C synthesis look-alike gene in the guinea pig or human DNA or we could admit the more obvious conclusion, that humans and primates and other mammals share a common ancestor.*[11]

Fourth, Johnson relies heavily on rhetoric. One rhetorical device that he uses to his advantage is the assumption that any disagreement between scientists on issues surrounding evolution is tantamount to saying that the entire model should be discarded. Thus he quotes extensively from Stephen Jay Gould and his idea of punctuated equilibrium as a way to deal with the sudden (in geological time) appearance of new forms in the fossil record. Despite the fact that Gould does not wish to throw out the Darwinian model in any of his writings, Johnson wrongly assumes that the disagreement is with the fundamental idea of the model, natural selection.

Finally, Johnson attempts to characterize natural selection and the survival of those most fit as a tautology:

> *The theory predicts that the fittest organisms will produce the most offspring, and it defines the fittest organisms as the ones which produce the most offspring.*[12]

We believe that Michael Ruse's refutation is adequate and effective.[13] Ruse makes three points that invalidate Johnson's objection. First, natural selection depends upon the fact that more offspring are produced than can survive. This is an observation that predates Darwin. Second, there are traits that differ between those members of the population that survive and those that do not. These trait differences can, in fact, be shown to be related to survivability. Finally, he notes that successful traits can also be shown by experiment to yield like results in like situations. These three observations mean that Johnson's simplistic tautology is actually an observational fact of some merit to the Darwinian model.

Turning now to the work of Michael Behe, we believe he has correctly noted the amazing structure of the intracellular systems that serve as his exemplars of irreducible complexity. However, Behe has made the error of the God-of-the-gaps theology. This theological position argues that anything which cannot routinely be explained by known physical laws must be ascribed to the intervention of God. In Behe's case, of course, it is an Intelligent Designer who assumes the role of filling in the gaps.

The problem with this position has been known for centuries. If we place God (or the transcendent Designer) within the gaps of our knowledge, then we run the risk that when science advances and a natural explanation is found, the place for God disappears. Modern theologians have frequently registered their doubts that such a God-of-the-gaps method is reliable for depicting God's action in the natural world. Is Behe's black box a Darwinian gap to be filled by God? Kenneth Miller, a cell biologist and theistic evolutionist at Brown University, has made just this critique of Behe. In his book *Finding Darwin's God,* Miller discusses each of the Behe exemplars of irreducible complexity and presents references that support a natural explanation rather than appealing to intelligent design. After reviewing a number of examples from the scientific literature that counter Behe's statements, Miller concludes:

> *Michael Behe's purported biochemical challenge to evolution rests on the assertion that Darwinian mechanisms are simply not adequate to explain the existence of complex biochemical machines. Not only is he wrong, he's wrong in a most spectacular way. The biochemical machines whose origins he finds so mysterious actually provide us with powerful and compelling examples of evolution in action. When we go to the trouble to open that black box, we find out once again that Darwin got it right.*[14]

Behe has responded to these criticisms and has engaged Miller in an active debate, both in person and on the Web. However, it still remains true that the logical result of finding Darwinian explanations for the evolution of Behe's examples is a theological pitfall of the argument. Behe unnecessarily confuses primary cause (divine action) and secondary causes (the lawlike behavior of the universe).

Of the three voices, that of William Dembski is the most challenging to critique. His argument relies on more than merely an analysis of the scientific claims of the Darwinian model. Unlike either Johnson or Behe, Dembski introduces a concept of his own—specified complexity. He does not reject the fossil record or the molecular evidence, as does Johnson. Like Behe, he argues that design is apparent in the structures found in the living world. And like Behe, Dembski offers what he calls a testable hypothesis: if specified complexity exists in nature, there are ways in which it can be detected.

The arguments Dembski raises are mathematical and philosophical rather than strictly biological. He approaches the questions from a radically different direction than those of either of the other two writers we have considered. He has been criticized vociferously in print and on the Web. He has responded at length to these critiques, both in print (*No Free Lunch*) and on the Web. These exchanges of views are ongoing and involve everything from the details of Dembski's mathematical arguments to the philosophical and theological positions maintained by each writer or commentator. We do not intend to review the entire structure of this discussion here. Instead, we wish to focus on the position taken by Dembski with reference to divine action.

While Dembski argues for the status of intelligent design as a scientific research program with testable hypotheses, he also deals with the question of who or what the intelligence is that is responsible for the design. There are two ways in which he treats this question. One is to argue for two kinds of causal explanations: intelligent causes and natural causes. The second is to argue that intelligent design, as a scientific research program, is not concerned with the nature of the designer.

In the first case, Dembski contrasts natural and intelligent causes throughout *No Free Lunch*. In the introductory material of this book, he is quite specific about what natural causes are:

> *Natural causes, as the scientific community understands them, are causes that operate according to deterministic and nondeterministic laws and that can be characterized in terms of chance, necessity, or their combination.*[15]

This definition is in concert with what we mean by secondary causes. However, what does Dembski mean by intelligent causes? Any causes that have telic or purposeful properties would constitute intelligent causes for Dembski. But is it not the case that local purpose belongs inherently to systems within the larger domain of secondary causation? Is it not obvious that the ocular system is organized for the purpose of making sight possible? Is the natural world after leaving the hand of the divine Creator not capable of self-organization, including the evolution of eyes? If so, then Dembski's argument is with the philosophical underpinning of science, not the scientific study of nature itself.

If Dembski means by "intelligent causes" (note his use of the plural) the direct intervention of God, then are we to assume that all of these causes are subsumed under what we, with Saint Thomas, refer to as the primary cause? Is Dembski saying that nature by itself is not capable of self-organizing, so it needs God to intervene to impart purposeful organization? If this is the case, then the discussion becomes theological and not scientific at all. To be scientific in our era is to search for solely natural explanations.

In fact, it is the theological aspects of Dembski's work that interests us in this analysis. One available model for divine action in nature is the front-loading concept, wherein the lawlike behavior of the universe is placed there from the beginning by God. God set the initial conditions, so to speak, at the Big Bang so that physical nature could develop life on its own over deep time. Dembski correctly suspects that this might be a deist position, where *deism* is the view that God the creator no longer acts in the world. However,

Dembski sees no other alternative to this than to invoke an interventionist position that is inconsistent with the methodology of science, in which natural processes have natural explanations (secondary causes). This places Dembski in the classic theist camp.

This leaves us with a split. It leaves us with a choice between a deist position where God refuses to act in nature on the one hand, and a classic theist position where God's interventions in nature are tantamount to miracles for advancing evolution on the other hand. Might there be a position in the middle? Might we think of ongoing divine action that is not interventionist? Yes. Consider the writings of Robert John Russell and William Stoeger. These two theistic evolutionists have been working on a noninterventionist yet objectivist concept of divine action in the natural world. Their assessments of how divine action can take place without compromising the methods or observations of science come from their understanding of the quantum view of the world.[16] Their ongoing research investigates how the quantum nature of the world, including such properties as quantum indeterminacy and quantum nonlocality, might allow a place for divine action without violation of the observed lawlike behavior of the universe. While they have not resolved all of the philosophical and theological issues that this raises, they have provided another way of considering divine action as a primary efficient cause and yet as a continuing force in ongoing creation (*creatio continua*). Our point here is that a noninterventionist model of God's ongoing activity in the natural world is available. By teaching the theological controversy, we would like to show the gaps in ID theory and provide a viable alternative.

It would appear then that Dembski's retreat into an interventionist deity and rejection of natural causes as "incomplete" may be premature.

What Constitutes the Best Science?

Finally, we feel obligated to say clearly why we believe intelligent design ought not be presented as an alternative to the theory of evolution in science classrooms. We believe that the Darwinian model with its recent updates in evolutionary biology should be made the standard, at least until a superior theory takes its place. Faith requires the best science, and to date the Darwinian model of evolutionary biology has proven to be the best model for generating progressive research.

The criterion that establishes the superiority of one model or theory over its competitors is fertility. By fertility we mean this: a theory or model of natural reality is fertile when it gives rise to a progressive research program, when it guides scientists in performing experiments that lead to new knowledge. The Darwinian model has proven itself fertile for a century and a half now.[17] The principles of random variation and natural selection gain repeated confirmation. Not only has it predicted what we would find when we examined the fossil record and the DNA record, but it has spawned the kind of knowledge that medical researchers need for producing therapies that eventually save lives. We are not saying that Darwinian theory is exhaustively or totally true; rather, we are saying that of all rival theories regarding life processes this one is the most progressive. At some later date it may be replaced by a more comprehensive paradigm; yet at this point in time, the Darwinian model constitutes the best science.

The model of nature put forth by intelligent design theorists cannot match this claim. The ID model is not fertile. No one expects it to generate progressive research protocols. No one expects medical science to employ ID for curing disease. It would be unethical to teach young people in our school classrooms any system of biology that fails to prepare them for the highest quality of scientific research they may be offered when they get to the university.

Religious Diversity and the Evolution Controversy

Is the controversy over the teaching of evolutionary biology in public and private schools just a conservative Christian idiosyncrasy? Is it restricted to North America, where fundamentalists and evangelicals have influence? Are intelligent design advocates merely evangelical missionaries in disguise? Do the other religions of the world go about their merry way without giving this controversy a second thought?

On the one hand, the debate over evolution in the schools is parochial. It concentrates primarily in Christian settings within North America or those influenced by North America. On the other hand, it spills well beyond its original borders. Intelligent design books have been translated into Chinese and are being read in Asia and elsewhere. The Web carries the debate to every part of our globe. Many religious leaders who might not have been originally interested are feeling a bit compelled to take a stand.

Among the scholars of religion, Huston Smith, author of the widely read book *The World's Religions*,[18] has come out in support of ID.[19] In contrast, Mary Evelyn Tucker, who directs the Forum on Religion and Ecology, sees the controversy over evolution as strictly a parochial matter that is limited to Christian theology in a North American context. It matters little if at all to the religious traditions in the rest of the world.[20]

Let us take a brief look at two other Abrahamic religious communions: Judaism and Islam. Judaism has almost fully reconciled itself to evolutionary theory, as it has to science in general. Islam is ambivalent, with supporters of evolution and detractors.

Judaism and Darwinism

The battle over evolution is somebody else's war, if you look at it through Jewish eyes. It is not a Jewish problem. According to Jewish theology, God's creation at the beginning was unfinished. So it is no surprise that a scientific theory such as evolution might arise that shows ongoing creation.

The history of reading the Bible through Jewish eyes is rich with interpretation. The tradition of *Halacha* permits and even encourages expansion from the tradition-specific character of God's revelation to the chosen people toward universal wisdom and shared knowledge. It permits and encourages healthy hospitality to science. The interpretation of Genesis is particularly instructive. Medieval theologian Moses Maimonides criticized those who insisted on reading Genesis literally. When the Torah appeared to conflict with what science would say, he would ask that we learn well what science would say. Then we would modify Torah interpretation accordingly. A conflict with the Darwinian model of evolution is less likely to arise in Judaism than in other religions of the Book.

Within half a century after the 1859 publication of the *Origin of Species*, the majority of Conservative and Reform Jewish intellectuals had come to accept Darwinism. Some Orthodox Jews down to the present see a conflict between the Darwinian model and allegiance to the biblical account of creation. Some Orthodox Jews find themselves in sympathy with Christian creationists. In summary, however, the dominant view of contemporary Judaism is to treat Genesis symbolically or figuratively, not literally; hence, relatively little difficulty with Darwinian evolution has arisen.[21]

Islam and Evolution

What stake, if any, does Islam have in the controversy over evolution? The answer is ambiguous. Some Muslims celebrate evolutionary theory, while others growl in suspicion and denounce it. Many are indifferent. For most Muslims, evolution belongs to the world of modern science, so it fits within the larger challenge of making peace with modernity.

"The global penetration of modern science is a *fait accompli*, whether one likes it or not," writes Muzaffar Iqbal, editor of the journal *Islam and Science*.[22] Some Muslims like it. They believe they can find precursors to modern evolution in their ancient religious roots. This leads Medhi Golshani in Tehran to review sacred passages, looking for anticipations of evolutionary thought. "God is the Creator of everything," says the Qur'an (13:16). Now, did God create once and for all, only at the beginning? Elsewhere we read: "Were we worn out by the first creation?" (50:15). Could this indicate a second creation or, better, continuous creation? Yes, answers Golshani.[23] If we find here an opening toward continuing creation, might this include the evolution of species? Might modern science add something that helps us interpret the Qur'an?

Let's return for a moment to the late nineteenth century when evolutionary theory first arrived on the scene. Lebanese scholar Husayn al-Jisr (1845–1909) argued that within Islam we can find honest attempts to reconcile modern science with classic faith. Like Christians worried about the relationship of an authoritative Bible with modern science, Muslims too need to find a way to interpret the Qur'an in light of the new understandings of science. One way is to declare that the new understandings are really not all that new. They were anticipated already in the divinely inspired Qur'an. This was al-Jisr's approach. He would quote passages from the Qur'an such as "we made every living thing from water," and then argue that this is reconcilable with the Darwinian model of evolution. After all, land animals and birds evolved from water life.

Al-Jisr quizzed himself regarding the origin of species. He could find no evidence within the Qur'an that we must assume all species were fixed at origin. They could have come into existence gradually. He found he could interpret the Qur'an to accommodate the new Darwinian model.[24]

"Today we know that Darwin's theory of evolution is actually the Muslim theory of evolution."[25] These are the words of Muslim T. O. Shanavas, who writes passionately to rescue Muslims tempted to sympathize with Christian fundamentalists, creationists, and evangelical supporters of ID. "Contrary to the current opposition to teaching evolution in American public schools, centuries before Darwin the doctrine of the gradual development of life forms ending in mankind was part of the curriculum in Muslim schools."[26] Premodern Islam could accept such things as deep time and human kinship with the apes, and contemporary Islam celebrates science and encourages the study of science. "Science and religion need not be competing ideologies," he says.[27]

So one wing of Islam is proevolution. However, there is another wing. Shiite scholar Seyyed Hossein Nasr argues that the premodern religious vision has its own integrity. It does not need to be propped up by science. "The religious view of the order of nature must be reasserted on the metaphysical, philosophical, cosmological, and scientific levels as a legitimate knowledge without necessarily denying modern scientific knowledge."[28] Although Nasr respects science, he advises religious persons in general, and Muslims in particular, to remain loyal to the fundamentals of the faith.

Like Phillip Johnson, Nasr takes a strong stand in opposition to the Darwin-inspired ideologies, especially atheistic materialism as promulgated by Herbert Spencer and Thomas Huxley. Nasr objects to the exhaustive materialism on the grounds that the religious vision retains the independent existence of mind, of spirit. The mind is an independent substance, says Nasr. It is a mistake to say that mind simply evolved from a material base. He also finds the expunging of purpose or design absurd, especially when evolutionists turn right around and embrace the pattern of natural selection. What is this pattern of natural selection if not a design? The idea of evolution is simply not good science, says Nasr; and it is unnecessarily destructive to religion. "The spread of evolutionism destroyed the very meaning of the sacredness of life and removed from nature any possibility of bearing the imprint of the immutable and the eternal."[29]

Other Muslims go farther down the antievolution road. Harun Yahya (the pen name of Adnan Oktar) in Turkey says Darwinian evolution is nonsense. "To believe that all the living things we see on Earth, the flowers with their matchless beauty, fruits, flavors, butterflies, gazelles, rabbits, panthers, birds, and billions of human beings with their different appearances, the cities built by these human beings, the buildings they construct, and bridges all came about by chance from a collection of mud, means taking leave of one's senses."[30] Then he pronounces the ultimate denunciation: evolution is Satanic. Satan is using the theory of evolution as a deceptive device to woo us away from Allah. Yahya finds alliances with Christian creationists helpful in his battle against Darwinian evolution, especially Darwinism as taught in the Turkish public schools. In Turkey, and spreading elsewhere in the Islamic world, we find an Islamic creationist movement.

In sum, Muslims wrestle with the degree to which evolutionary biology should be accepted or embraced. On the one hand, Muslims who admire the modern world find ways to interpret the Qur'an so that synthesis can result. On the other hand, Muslims who are engaged in their own culture war between Islam and secularism are willing to borrow the weapons crafted by North American creationists and intelligent design theorists. What the future holds here is contingent and unpredictable.

Conclusion

We have attempted here to provide a theological and religious analysis along with a scientific evaluation of intelligent design. It is our considered judgment that intelligent design does not qualify as a fertile or progressive scientific theory, so it ought not to be taught to public or religious school children in science classes.

As we indicated earlier, our commitment to the best science is a commitment that arises out of our Christian faith. Jesus says in John 8:32 (New Revised Standard Version), "You will know the truth, and the truth will make you free." All truth comes ultimately from God, we along with many of our Jewish and Muslim colleagues believe. And this includes mundane scientific truth as well as sublime revelation.

The Darwinian model may not be the whole truth, to be sure. No scientific theory can count as apodictic truth. Yet there is enough truth in evolutionary theory to spawn progressive research and advance the study of medicine and related disciplines. This makes it the best science available today. We believe it is our ethical responsibility to provide young people in classrooms with the best science available, and today that's the Darwinian model of evolutionary biology.

4: From the Classroom to the Courtroom: Intelligent Design and the Constitution

JAY D. WEXLER

Teaching intelligent design in our nation's public school science classrooms is no doubt bad educational policy, but is it also unconstitutional? The only court to consider the issue so far answered with a resounding yes,[1] but since its decision lacks precedential authority outside a very small area in Pennsylvania, the question is still a live one. It is hard to say with any certainty what some other court might do if faced with another challenge to an intelligent design policy. The Supreme Court's religion law jurisprudence is notoriously vague and indeterminate, and a lot will necessarily turn on the specific circumstances of whatever case or cases find their way to the courts. Despite these uncertainties, however, it is quite clear that teaching intelligent design in public science classrooms would raise serious constitutional problems in most if not all cases, and that schools that adopt intelligent design policies run a substantial risk of losing in the courts. This chapter will canvass the legal issues relevant to the adoption of an intelligent design policy in the public schools, and it will suggest that such policies, in most cases, should be found unconstitutional.[2]

Church-State Relations and the First Amendment

The first sentence of the First Amendment to the U.S. Constitution says "Congress shall make no law respecting an establishment of religion or prohibiting the free exercise thereof." The first part of the sentence is usually called the Establishment Clause; the second part is known as the Free Exercise Clause. Although the sentence speaks only in terms of what Congress may or may not do, the Supreme Court has long held that the First Amendment in fact applies to all levels of government, including state and local bodies. Like many parts of the Constitution, the religion clauses of the First Amendment can be interpreted in many ways, and so the courts have had to struggle over the past century to clarify what kinds of actions government can and cannot take with respect to religion. It would be impossible in the space provided here to give any kind of a comprehensive summary of the history of church-state law, so I will limit myself to a very brief sketch of the current law regarding the establishment of religion, with specific emphasis on the Supreme Court decisions that have set out the legal tests most relevant for the constitutional evaluation of any policy involving intelligent design.

The most general statement of Establishment Clause law is the so-called *Lemon* test. According to this test, first articulated by the Supreme Court in a case called *Lemon v. Kurtzman,*[3] the government may not take any action that lacks a secular purpose, has the primary effect of advancing religion, or fosters an excessive entanglement between religion and the state. This general test, however, creates more questions than it answers. For example: How do we determine the government's purpose? What constitutes an advancement of religion? How much entanglement is excessive? And what counts as entanglement anyway? As a result, the Court rarely looks only to this test to decide cases. Instead, the Court applies more specific variations or elaborations of the *Lemon* test depending on what kind of challenge it is considering.

One such elaboration that is particularly relevant for considering the constitutionality of intelligent design policies is the endorsement test. Created by Justice Sandra Day O'Connor in a concurring opinion in 1984 and adopted by five members of the Court five years later in a case involving various holiday displays near a Pittsburgh courthouse, the test asks whether a "reasonable observer" would feel that the government has sent a "message to non-adherents that they are outsiders, not full members of the political community, and an accompanying message to adherents that they are insiders, favored members of the political community."[4] In determining whether the government has violated the test, the Court considers the entire context of the government's action, particularly the action's historical context, and measures the perception of "the reasonable observer," who "must be deemed aware of the history and context of the community and forum in which the religious display appears."[5]

The Court has twice struck down attempts by states to interfere with evolution education. In 1968, in a case called *Epperson v. Arkansas*, the Court invalidated an Arkansas law that prohibited teachers from teaching evolution, finding that the law was motivated by a completely religious purpose.[6] About twenty years later, the Court held that Louisiana's equal-time law, requiring schools to teach so-called creation science whenever presenting evolution, also violated the Establishment Clause. This decision, called *Edwards v. Aguillard*,[7] is the most relevant statement by the Court to date for considering whether a public school can constitutionally introduce the concept of intelligent design into science classes.

As a formal matter, the Court in *Edwards* decided against the statute for the same reason it invalidated the statute at issue in *Epperson*: It found that the statute lacked any secular purpose. But because the state had in fact claimed a secular purpose—promoting the academic freedom of public school teachers—the Court engaged in extended analysis on the way to concluding that this claimed purpose was in fact a sham. In the course of this discussion, the Court relied on several key facts and findings: (1) the poor means-end relationship of the statute, that is, the poor fit between the goal of promoting academic freedom and what the statute actually did; (2) the historic link between religion and critiques of evolution; (3) the singling out of evolution from among all possible topics in the curriculum for reform; (4) the favoring of creation science under certain provisions of the statute; and (5) statements from the legislative history indicating the legislature acted with the intent to promote religion.

Because the Court found that the statute was unconstitutional on *Lemon*'s first prong, it did not also consider whether the statute had the unconstitutional effect of endorsing or promoting religion. But given the Court's skeptical attitude and the fact that the five factors examined by the Court probably equally supported a finding of endorsement or promotion, the Court most likely would have struck down the statute on these grounds as well, had it considered them. Since it found that the legislature had acted with a religious purpose to promote and endorse religion, the Court could only have found a lack of actual endorsement or promotion if it had found that the legislature had not in fact accomplished its goals by issuing the law, a conclusion that seems quite unlikely given the Court's attitude toward the statute.

Finally, it is worth noting that the Supreme Court has been particularly vigilant in policing the Establishment Clause within the public schools. Specifically, in the eight cases in which the Court assessed the merits of an Establishment Clause challenge to an arguably religious practice (school prayer, moment of silence, and so on) in a public school, it has struck that practice down as unconstitutional.[8] The Court has given a variety of reasons for strictly applying the Establishment Clause in the public school context, including the compulsory nature of public education, the authoritative role that teachers play in the classroom, and the impressionable nature of young children and adolescents. As a result, any attempt to introduce intelligent design into the public school science classroom will have to pass a high hurdle to survive constitutional challenge.

Intelligent Design: Is It Religion?

Logically, the first issue to consider is whether intelligent design is religion, as that term is used in the First Amendment. If it is, then schools may not advance, endorse, or promote it in the classroom.

Perhaps surprisingly, the Supreme Court has never defined the term *religion* in the First Amendment, although it has hinted in some nonbinding language that the term should be understood quite broadly to encompass more than traditional Judeo-Christian monotheism.[9] Lower courts struggling to give the term some meaning have adopted a test that asks whether the belief system in question is a comprehensive viewpoint that addresses fundamental questions and is associated with certain formal and external signs common to most religions, such as holidays, symbols, and professional clergy.[10] Defenders of intelligent design have argued that under this framework intelligent design is not a religion; they argue that intelligent design lacks the external attributes of most religions, is an isolated teaching rather than a comprehensive one, and does not address "fundamental and ultimate questions having to do with deep and imponderable matters."[11] Although these writers may underestimate the extent to which intelligent design addresses fundamental questions, they are probably on strong ground to conclude that under this particular test, intelligent design does not constitute a religion.

This test, however, cannot be the right test for evaluating the constitutionality of intelligent design. If it were, then schools could also encourage students to pray, since the concept of prayer, by itself, does not meet the requirements of the three-part test either. Likewise, if the analysis of the intelligent design supporters were correct, then schools could teach the truth of karma, sin, reincarnation, or any other indisputably religious concepts because none of these concepts by itself would meet the three-part test. These obvious examples demonstrate that courts must apply a different test when the issue is whether some concept, practice, or belief in isolation is religious, as opposed to whether some integrated belief system counts as a religious belief as a whole.

The courts have not explicitly recognized this problem, but the right analysis for the question would ask whether the concept, practice, or belief in question resonates in religion rather than in some other area of inquiry, so that a reasonable person would understand the government promotion of the concept as a promotion or endorsement of religion. Reasonable inquiries under such a test would include whether an average person would associate the concept primarily with religion; whether the concept is an important aspect of the religious traditions that people are generally familiar with; whether people also commonly associate the concept with ideas or belief systems that are not generally viewed as religious; and whether, if the concept is associated with nonreligious belief systems, it is more prominently associated with those belief systems than with religious traditions.

Although application of this test might be difficult for some potentially religious concepts, intelligent design is in fact quite easy to analyze under the framework. Does intelligent design resonate in religion? Does the notion that an intelligent designer created the world and all of its inhabitants resonate in religion? Of course it does. The intelligent design of the universe is the core concept of the major prominent Western religions, without which those religious traditions would be unrecognizable. Most reasonable people would associate the intelligent design of the universe with religion. No significant nonreligious school of thought has an intelligent designer or creator as a core concept. And the Supreme Court in *Edwards* specifically characterized as a "religious viewpoint" the belief that "a supernatural being created mankind." Thus, it seems clear that intelligent design should be considered a religion for First Amendment purposes. However, whether intelligent design constitutes a religion may be largely beside the point. After all, although public schools cannot promote or advance or endorse or teach the truth of any religion, they are perfectly free to teach *about* religion. They can teach *about* Christianity, *about* Judaism, *about* Hinduism, and *about* Confucianism, and there are many good reasons for schools to teach these subjects.[12] So, if public schools can teach about religion, why shouldn't they be able to teach about intelligent design? To some degree they certainly can. For example, if a public school chose to teach about the philosophical claims of intelligent design in a philosophy of science class, the intelligent design movement in a current affairs class, or about the truth claims of intelligent design in a comparative religion class, most likely these choices would not pose a constitutional problem, assuming that intelligent design is not presented as a scientifically credible alternative to evolution.

Things are very different, however, when schools propose to teach intelligent design as a legitimate alternative scientific theory to evolution. For one thing, because science teachers generally present the best thinking in the field as the current state of knowledge without pointing out each and every dissenting view, even well-intentioned teachers trying to paint a balanced picture may end up leaving students with the impression that intelligent design is in fact true. This problem is exacerbated by the lack of adequate materials for teachers to use to teach evolution and intelligent design together in an objective fashion. Most important, even a policy that urges schools to use an objective approach to teaching about intelligent design might constitute an unconstitutional endorsement of religion or be motivated by a predominantly religious purpose, so that the very adoption of the policy would be unconstitutional. Whether this would be the case turns in large part on the proper understanding of the Supreme Court's decision in *Edwards*, to which the chapter now turns.

Intelligent Design and the *Edwards* Case

Recall that in striking down Louisiana's equal-time-for-creation-science law, the Supreme Court looked to a number of factors in determining that the legislature's articulated secular purpose was a sham. Among those factors were the poor fit between that purported goal and the actual language of the statute, the singling out of evolution for special treatment, the long history of religious opposition to evolution, and statements from the law's sponsors indicating an intent to promote religion. As it turns out, the intelligent design movement shares many of the same problems that doomed the creation science movement nearly twenty years ago. Stated strongly, the Establishment Clause case against intelligent design can be summarized in terms of the *Edwards* factors as follows: Against a long and visible history of clearly religious opposition to teaching evolution, once more a movement arrives that speaks in explicitly religious terms and singles out evolution from among all topics in the school curriculum for change, with the stated goal of informing students about a significant scientific controversy which in fact does not exist. What message would a school send to a reasonable observer by embracing such a movement? Most likely, the reasonable observer would understand that the government has reformed the curriculum for religious reasons, which is precisely what the Court in *Edwards* said the government may not do. The following discussion focuses on three of intelligent design's most damaging constitutional problems: its singling out of evolution education for reform, its explicitly religious background, and its status as unsuccessful science.

First, like the iterations before it, the intelligent design movement singles out evolution as the one subject in the entire curriculum that deserves criticism. This clearly demonstrates that the real intention of the movement is not to promote any secular goal but rather to promote a religious viewpoint, and it also ensures that the message received by reasonable observers will be the same. To understand this better, consider what a policy that truly sought to promote real secular goals would look like. For example, if reformers sincerely intended to make sure that students understand minority views on science, or the process of science as a discipline that is in constant flux as new discoveries are made and explanations developed, then why would those reformers focus exclusively on evolution rather than on subjects throughout the science curriculum? Indeed, several real-world proposals exist to teach students about the scientific process and minority views across various scientific disciplines,[13] but intelligent design supporters do not appear to be rushing to support these efforts.

Likewise, although intelligent design proponents often complain that the public school curriculum is not neutral with respect to religious views, a complaint that is at the same time both true and inevitable, they surely are not proposing to reform the curriculum so that it is completely neutral with regard to every religious belief held by anyone in the community. For example, some religious traditions teach that the world was created by numerous deities or by an animal such as a turtle or a raven, often in a very specific fashion that does not involve the simple creation of life by an intelligent agent.[14] Teaching intelligent design is no more neutral with regard to these specific creation beliefs than teaching evolution is, but one does not find intelligent design advocates suggesting that public schools should teach a wide variety of creation stories in order to maintain neutrality in the public school curriculum.

The troubling focus on singling out evolution for reform signals an important potential constitutional problem for the emerging arguments-against-evolution movement, which may, in the aftermath of *Kitzmiller,* become a major battleground for future evolution wars. Although a full constitutional analysis of this movement falls beyond the scope of this chapter, it is worth pointing out that, as with the creation science and intelligent design movements, the notion that schools should teach minority critiques of only one specific scientific theory rather than all similarly situated theories is vastly underinclusive with regard to any possible secular goal that could be articulated to support it. As a result, anyone trying to understand why some public school system would make a conscious decision to teach widely rejected criticisms of only evolution (and not, say, gravity, or the roundness of the earth, or some other widely accepted scientific notion) would surely surmise that the reform was intended to promote the clearly religious view that evolution is an unconvincing scientific position.

Second, as with past efforts to discredit evolution, the intelligent design movement often speaks in explicitly religious terms, thus compounding the religious message that any reasonable observer would take from the attempt to adopt an intelligent design policy in the public schools. There are two general ways that the movement can be understood as speaking in religious terms. First, the founders and leaders of the movement—people such as Phillip Johnson and William Dembski and organizations such as the Discovery Institute—often talk generally about intelligent design using religious language. Second, the political decision-makers who make the actual decision to implement some particular intelligent design policy in a specific public school or school system may also speak in explicitly religious terms during the lead-up to implementation, as well as its follow-through.

Much more can be said in concrete terms about the first of these than the second. As the federal judge in *Kitzmiller* rightly pointed out, many of the most prominent individual and institutional supporters and developers of intelligent design theory have consistently spoken about that theory in religious terms. Perhaps most damning on this score is the so-called Wedge document, which stated one goal of the intelligent design movement as "replac[ing] materialistic explanations with the theistic understanding that nature and human beings are created by God." And as made clear by Barbara Forrest at the trial, the various drafts of the primary intelligent design textbook *Of Pandas and People* demonstrate that the authors simply replaced the terms *creation* and *creation science* with *intelligent design* to avoid running afoul of the Supreme Court's decision in *Edwards*.[15]

Less can be said about the language used during the deliberations over any specific intelligent design policy because that language will differ depending on the circumstances. It is at least theoretically possible that a school could adopt an intelligent design policy without talking in religious terms at all. For example, an individual science teacher could introduce the topic into a classroom without mentioning religion, assuming there is no contrary school policy in place. If there is a scenario for introducing intelligent design in a science class that could possibly pass constitutional muster (and even this is probably unlikely), it would involve such an individual teacher who sincerely thinks intelligent design is an interesting scientific concept for reasons truly unrelated to religion and says exactly that upon introducing the subject. On the other hand, it seems far more probable that most decisions to adopt intelligent design would resemble the situation in Dover, Pennsylvania, where supporters and school board members spoke in favor of the policy in explicitly religious terms, with one member saying something like, "Two thousand years ago someone died on a cross. Can't someone take a stand for him?" Language like this quite obviously will weigh in favor of finding any specific intelligent design policy unconstitutional.

It is important when analyzing religious language that is used in support of some government policy or legislation that courts be careful not to place too high a bar on the use of religious language during the policy-making or legislative process. Like all individuals, government officials possess a First Amendment right to speak in whatever terms they wish about a given policy in the political arena, and a liberal democracy that treats religious citizens with respect and equal regard should not censor too strongly the invocation of religious beliefs that happen to coincide with some public policy decision.[16] A law should not find itself in constitutional jeopardy simply because some citizens speak in its favor in religious terms; if that were the law, neither the abolitionist nor the civil rights movement would have fared well in the courts. Religious statements in support of a challenged government policy or law become problematic only when they so suffuse the decision-making process that they demonstrate that the law was promulgated predominantly for a religious purpose or they contribute to sending a message to the reasonable observer that the government is endorsing a religious belief. This is what happened in the Dover case.

Finally, and perhaps most important, as with the creation science equal-time law at issue in *Edwards*, the intelligent design movement suffers from an enormous disconnect between the movement's means and its purported ends. To the extent that the movement argues that public schools should teach intelligent design in science classes to inform students of an important scientific controversy, it fails, because there is no such controversy. As other chapters here have explained in detail, intelligent design has found no success in the scientific community. That community, including almost every major scientific organization, has universally accepted evolution as a central theory in biology and roundly rejected the concept of intelligent design as an alternative to evolution. Most telling, perhaps, is intelligent design's near total failure to make any headway in the peer-reviewed publications that are the gateway to scientific success. With this justification for teaching intelligent design completely undermined by the facts, what possible reason could there be to teach it in science classes, other than to discredit a theory that many religious people find untrue and offensive? As a result, the message sent by any school that adopts an intelligent design policy will almost surely be that it intends to promote the religious belief that an Intelligent Designer created the universe.

The *Kitzmiller* Decision

As of this writing, the issue of intelligent design's constitutionality has reached the courts only one time. In late December 2005, Judge John E. Jones III, a United States district court judge sitting in Harrisburg, Pennsylvania, issued a 139-page ruling striking down as unconstitutional an intelligent design policy adopted by the Dover Area School Board in Dover, Pennsylvania, the previous year. The challenged policy included two parts. First, the board had adopted a resolution stating: "Students will be made aware of gaps/problems in Darwin's theory and of other theories of evolution including, but not limited to, intelligent design." Second, the board had issued a statement, required to be read to ninth-grade biology students in the district, which said in full:

> *The Pennsylvania Academic Standards require students to learn about Darwin's Theory of Evolution and eventually to take a standardized test of which evolution is a part.*
>
> *Because Darwin's Theory is a theory, it continues to be tested as new evidence is discovered. The Theory is not a fact. Gaps in the Theory exist for which there is no evidence. A theory is defined as a well-tested explanation that unifies a broad range of observations.*

> *Intelligent Design is an explanation of the origin of life that differs from Darwin's view. The reference book,* Of Pandas and People, *is available for students who might be interested in gaining an understanding of what Intelligent Design actually involves.*
>
> *With respect to any theory, students are encouraged to keep an open mind. The school leaves the discussion of the Origins of Life to individual students and their families. As a Standards-driven district, class instruction focuses upon preparing students to achieve proficiency on Standards-based assessments.*

After various parents of children in the Dover school system sued to enjoin the policy, Judge Jones conducted a trial over the course of six weeks to consider whether the policy violated the Establishment Clause. His comprehensive opinion in the case of *Kitzmiller v. Dover Area School District,* finding the school board's intelligent design policy unconstitutional, was widely hailed as a tremendous victory for defenders of evolution.

Judge Jones's legal analysis proceeded in six logical steps. First, the judge concluded, based on his analysis of precedent from both the Third Circuit Court of Appeals (the appellate court that reviews his decisions) and the Supreme Court, that he would apply both the *Lemon* test and the endorsement test to the school board's actions. Second, he analyzed the historical development of the intelligent design movement, including its roots in medieval theology and creation science, as well as the religious language invoked by its supporters, and concluded that an objective observer would understand intelligent design to be a religious strategy "that evolved from earlier forms of creationism." Third, he considered how a reasonable student in the Dover schools would view the policy. Looking at the impressionable nature of the students, the specific language of the disclaimer, and the circumstances surrounding the classroom presentation of the disclaimer, the judge found that "an objective student would view the disclaimer as a strong official endorsement of religion or a religious viewpoint." Fourth, he concluded that reasonable adults in the Dover community would reach the same conclusion, because the board defended the policy in a very public fashion through community meetings and a widely circulated newsletter that described the policy in expressly religious terms.

Although Judge Jones could have ended his analysis at this point, he nonetheless proceeded to consider two other issues that he deemed important to evaluating the constitutionality of the board's policy. Thus, the opinion's fifth conclusion was that intelligent design is not science. The judge rested this determination on his judgment that intelligent design fails to follow the ground rules of science by relying on supernatural explanations, that it uses "the same flawed and illogical contrived dualism that doomed creation science in the 1980's," and that it has been completely rejected by the scientific community. Finally, observing that "the better practice in this Circuit is for this Court to also evaluate the challenged conduct separately under the *Lemon* test," Judge Jones ended his opinion by finding that the Dover policy independently failed to pass constitutional muster because the school board that enacted the policy was animated by the primary purpose of advancing religion. On this score, the judge canvassed with painstaking detail the events leading up to the adoption of the school board's policy—including various public meetings and the donation of sixty copies of the intelligent design textbook *Of Pandas and People* to the school district—and concluded that the board's purported secular purposes were in fact a "sham."

It is somewhat difficult at this time to assess the future impact of the *Kitzmiller* decision on the intelligent design movement as a whole. On the one hand, the decision has precedential value in only one small district in central Pennsylvania and therefore does not prohibit another district in some other area of the country from experimenting with intelligent design in its public schools. Thus, if a school district in Kansas or Ohio or Utah wants to introduce intelligent design into its science classrooms, *Kitzmiller* erects no legal barrier. On the other hand, the court's careful and detailed fact-finding, along with its persuasive and impressively complete legal analysis, will likely make school districts around the nation think twice (at least) about whether bringing intelligent design into their biology classes would really be worth the risks.

Judge Jones's opinion is too comprehensive to provide any type of exhaustive analysis here, but it is worth pausing to emphasize a few of the more important aspects of the decision. For one thing, Judge Jones appropriately spent most of his effort in the opinion applying the endorsement test to the school board's policy, a move that is not obviously dictated by the Supreme Court's case law. Although the Court has invoked the endorsement test in all sorts of cases involving challenges to many types of government action, including school funding and school prayer, the test is most relevant and useful to evaluating the constitutionality of religious displays, such as crèches, crosses, and the like, which send the message that government is favoring religion. Thus, the board argued in the case that the endorsement test was inapplicable to a curricular policy like the one at issue, especially given that the Court did not apply the test in either *Epperson* or *Edwards*, the two evolution cases previously to reach the Court.

Judge Jones's response to this argument was typically comprehensive and compelling. Not only did the judge make the obvious point that *Edwards* and *Epperson* predated the Court's adoption of the endorsement test, he also noted that *Edwards* in fact did mention the endorsement idea; that the Court, having ruled against the government on purpose grounds in that case, would probably not have applied the endorsement test even if the test had existed at the time; and that both the Supreme Court and the Third Circuit Court of Appeals have consistently applied the endorsement test in cases not involving religious displays. But perhaps most important, the judge understood that the endorsement test is the test best suited to evaluating a policy such as the one adopted by the Dover school board, because such a policy is problematic mostly due to the fact that it sends the harmful *message* that the government wants to change the curriculum to place a stamp of approval on a particular religious belief. The question of whether the school board had endorsed a religious message by adopting its intelligent design policy was therefore the central issue in the case, and Judge Jones rightly recognized it as such.

Of course, Judge Jones did not stop writing after finishing his endorsement analysis, and his decision to consider the purpose inquiry under the *Lemon* test as well was insightful and shrewd. Not only does the application of both tests make the decision more difficult to overturn in the unlikely case of an appeal (since the reviewing court would have to find the judge's analysis incorrect on both counts), but it also insulates the decision in the not entirely unlikely case that the newly constituted Supreme Court does away with the endorsement test. With the test holding on by only one vote and with Justice O'Connor, the test's creator and greatest champion, stepping down from the Court, it is certainly possible that the test, which has been the object of substantial judicial and scholarly criticism,[17] will be discarded by a more conservative Court. If Judge Jones had not decided the case on both endorsement and purpose grounds, a decision by the Supreme Court to get rid of the endorsement test would leave the *Kitzmiller* decision on shaky ground indeed. This way, the Court would have to get rid of both the endorsement and the *Lemon* test for the judge's decision to lose force. Although this too is not impossible, as the *Lemon* test has also been widely criticized, it is far more unlikely that the Court would reject both prevailing tests than just one.

Finally, it should be of no surprise to anyone familiar with reading legal opinions that one of the judge's more astute and important conclusions can be found in a footnote. At trial, the school board had argued "vigorously" that the act of reading the evolution disclaimer in front of the classroom did not constitute the teaching of intelligent design but simply made the "students aware of it." As such, the board argued, the policy could not have violated the Establishment Clause. The judge rightly rejected this argument with a two-part response. First, he argued that the Establishment Clause not only prohibits the teaching of religion, it also forbids the government from endorsing or advancing religion generally, which can be accomplished by actions in the classroom short of teaching. But more fundamentally, the judge also accepted the argument, advanced by various teachers at the trial, that reading the disclaimer is in fact teaching. As the judge found, "An educator reading the disclaimer is engaged in teaching, even if it is colossally bad teaching ... The disclaimer is a 'mini-lecture' providing substantive misconceptions about the nature of science, evolution, and ID which 'facilitates learning.'"[18] This notion that given the teacher's authoritative position in the classroom nearly everything he or she does there is in some real and significant sense "teaching" is perhaps a subtle point, but it is surely true and potentially an important legal conclusion. Like so much in the opinion, this footnote demonstrates that Judge Jones had an impressively firm grasp not only on the formal legal issues raised by the school board's flawed policy, but on the real-world stakes it involved as well.

Academic Freedom

Intelligent design advocates often argue that public school teachers have a First Amendment academic freedom right to teach intelligent design even if the school or school board has prohibited teachers from discussing the topic. For example, Francis Beckwith has argued that "bringing into the classroom relevant material that is supplementary to the curriculum (and not a violation of any other legal duties), when the public school teacher has adequately fulfilled all of her curricular obligations, *is protected speech* under the rubric of academic freedom."[19] According to another intelligent design supporter, "Public school teachers are protected in their classroom discussions by free speech and academic inquiry rights under the First Amendment speech rights."[20] In support of this argument, intelligent design advocates often cite language from *Keyhishian v. Board of Regents,* where the Supreme Court opined that the First Amendment "does not tolerate laws that cast a pall of orthodoxy over the classroom."[21] In light of the rather devastating opinion in *Kitzmiller,* perhaps intelligent design advocates will turn away from promoting districtwide policies mandating the teaching of intelligent design and move instead toward encouraging individual teachers to bring up the subject on their own. But what if the school has a policy (or subsequently adopts a policy) prohibiting such instruction? Would the teacher have the right to teach intelligent design anyway?

Intelligent design advocates are certainly correct to claim that the First Amendment places some limits on the state's authority to fire government employees, including public school teachers, and that those teachers do not completely give up their constitutional right to free speech when they accept a government job. The Supreme Court, in *Pickering v. Board of Education,* held that public school teachers have a limited right (subject to a balancing test, in which the court weighs the interest of the speaker against the countervailing government interests) to speak as citizens on matters of public concern without having to fear that their employers will fire them or otherwise take negative employment action against them.[22] But this right to speak out as citizens, in forums such as newspaper editorials or public meetings, is completely different from the asserted right to include material or views in the classroom in violation of orders from the government-controlled supervisory body. This latter right simply does not exist. It is unsupported by Supreme Court precedent, would undermine the democratic accountability of public schools, and would cause chaos in the nation's educational system.

For one thing, the Court has never held that public secondary school teachers have any independent "academic freedom" right. The words *academic freedom* do not appear anywhere in the Constitution, and although the phrase can occasionally be found in nonbinding language from the Supreme Court, that Court has never relied upon an academic freedom rationale to invalidate any government law or practice. As the Fourth Circuit Court of Appeals recently explained, "The Supreme Court has never set aside a state regulation on the basis that it infringed a First Amendment right to academic freedom. . . . [T]o the extent it has constitutionalized a right of academic freedom at all, [the Court] appears to have recognized only an institutional right of self-governance in academic affairs."[23]

Second, the lower courts that have recently considered the issue have generally rejected the notion that government employees have any First Amendment right to speak *in their role as employees* in violation of the dictates of their democratically accountable supervisors. This means that even if a teacher may have a right to speak out as a private citizen in favor of intelligent design (or unprotected sex or communism or any other unpopular idea) in a public meeting or newspaper editorial without fear of being fired, the same teacher does not have the same right to advocate those ideas inside the classroom, if the school (or school board or state) has provided clear notice that the teacher may not teach those subjects, or otherwise provided the teacher with adequate due process protections. Again, the Fourth Circuit made this clear when it explained that before applying the *Pickering* balancing test to a public employee, the court must decide whether the employee is speaking as a citizen or as an employee: "This focus on the capacity of the speaker recognizes the basic truth that speech by public employees undertaken in the course of their job duties will frequently involve matters of vital concern to the public, without giving those employees a First Amendment right to dictate to the state how they will do their jobs."[24]

This analysis is entirely sensible. If high-level government officials cannot restrict the official speech of their employees, then those employees (including teachers) would have near-complete authority to countermand the state's official messages. The Fourth Circuit invokes the example of a formal press conference where an assistant district attorney criticizes his boss's decision to pursue a murder charge, but one can imagine an endless stream of analogous examples where adopting this academic freedom argument would disrupt the government's functioning. Should the president's press secretary be free to criticize the administration's social policy? Should a scientist from the Environmental Protection Agency be able to officially state that some type of pollution is far less dangerous than the agency has recognized? Should a state employment officer be able to officially speak out against the state's prolabor policies? These examples demonstrate the chaotic results of recognizing a First Amendment right for a subordinate to speak in his or her official capacity on matters of public concern.

Recognizing a strong First Amendment speech right for state employees acting as employees would be just as problematic in the public schools as it would be anywhere else in the government. Intelligent design supporters do not propose any principled way to limit their academic freedom argument to the intelligent design context, which is unsurprising, since no principled limit exists. As a result, teachers could teach their personal views on a whole smorgasbord of controversial topics, and the school would be unable to stop them. Teachers could supplement a sex education class with their own views about how HIV is *really* transferred, suggest that the federally funded abstinence lesson they just taught is a "bunch of crapola," mention at the end of their health lesson that drugs are in fact "kind of a blast," hint that the horrors of the Holocaust have been a "bit overstated," or argue that slavery was a mutually beneficial economic arrangement for blacks and whites alike.

Allowing government supervisors to control the official statements of their subordinates ensures that government decisionmakers remain democratically accountable for the official messages of the state. Those who speak on the state's behalf are ultimately speaking for its citizens, and those citizens ought to be able to take some action if the state decides to take an official position that the citizens disagree with or find offensive. The electoral process gives citizens this power, but only for the highest level officials. From a democratic standpoint, then, it makes sense that those highest level officials ought to have the final say with respect to what messages the state will espouse. If the courts adopted the position urged by supporters of intelligent design, then citizens would lose any real power to hold the government accountable for its statements in cases where an employee makes an official statement on a controversial subject that contradicts the state's own message. So, if a public school teacher decides to teach that the Holocaust never happened, the community should have the power to pressure the school board to prohibit the teacher from advancing this view in the classroom. If the board can control the teacher's speech, and the board agrees with the community, then the board can fire the teacher if the teacher persists in teaching the objectionable viewpoint. If the board, on the other hand, decides not to reprimand the teacher, the community can remove the relevant members from the board at the next election. But if the teacher possesses a First Amendment right to say what he wants, there will be nothing that the community can do to stop the teacher from continuing to engage in the unwanted speech.

Some intelligent design supporters, such as Beckwith, argue only that a teacher has the right to *supplement* the existing evolution curriculum with intelligent design theory, rather than that a teacher has a First Amendment right to *replace* the prescribed curriculum by teaching intelligent design theory instead of evolution.[25] This distinction makes no difference, however. There is no reason to think that the analysis should be any different just because the employee first says what she is supposed to say before putting forth her own opinion. A First Amendment rule allowing supplementation of the curriculum but not replacement of curricular content would still undermine the functioning of government and obstruct the lines of democratic accountability. Should the president's press secretary really be able to officially state that "the president believes his social policy is just, but he is sadly mistaken"? Should the First Amendment really protect a public school teacher who wants to tell his class, "Most people believe that the Holocaust happened, but I have to say that the evidence looks a little thin to me" or "The principal thinks you shouldn't have sex, but I think he's sort of overreacting" if the community strongly disagrees with these statements?

Finally, it is worth emphasizing that nothing said here is intended to suggest that schools should generally restrict what their teachers may say in the classroom. Strong educational policy arguments surely exist for allowing teachers wide latitude to teach the material they wish in whatever way they want, even if sometimes their methods or materials may be controversial. Providing teachers this freedom also sends the essential message that teachers are respected professionals whose work is vital to the effective functioning of a democracy and the well-being of its citizens. Indeed, probably in almost all cases, school boards should allow teachers great leeway to do what they want when they are in front of their classes. This conclusion, however, is based on sound educational policy, not constitutional law. Taking the position that schools should usually allow teachers to supplement the curriculum with their own views is far different from saying that teachers have a First Amendment right to supplement the curriculum with their own controversial viewpoints in those isolated cases where the community strongly opposes that viewpoint. The latter purported right is unsupported by constitutional text or precedent and is contrary to common sense and the ideals of the political community.

Conclusion

The current dispute over teaching intelligent design in public school classrooms is the most recent chapter of the long-standing controversy over how schools ought to present the theory of evolution in their science classes. From time to time, democratically elected majorities have succeeded in implementing policies intended to undermine the teaching of solid science. Each time this has happened, however, these classroom initiatives have failed in the nation's courtrooms. The judges sitting in these courtrooms have uniformly realized that given evolution's commanding support in the scientific community, attempts to weaken evolution's status in the classroom are constitutionally suspect. As long as evolution continues to hold the dominant position in the scientific community that it has occupied for most of the past century, courts will likely continue to look at attempts to discredit it with similar disfavor.

5: Evolution in the Classroom

BRIAN ALTERS

> *I don't think that any job in the entire world—and I include Popes, Presidents and Generals—could possibly be more important than teaching science to secondary school students.*
>
> STEPHEN JAY GOULD, from an April 3, 1997, letter to McGill University

What is taught in high schools about biological evolution, and what should be taught about it? These questions must be tremendously important—after all, U.S. presidents, senators, and governors have recently voiced strong opinions on the subject of teaching about evolution, as have a number of religious leaders. The media coverage on this topic has been enormous, especially for a high school science curricular debate.

Federal and local court decisions, school board policies, administrative directives, and the positions of leading scientific and science education associations are all very important influences on what goes on in the classroom. However, as we educators know well, when the classroom door closes and the class begins, it is the individual teacher who has the greatest effect on what students learn, for better or for worse. Unfortunately, it is perfectly possible for a teacher to cover evolution in a way that satisfies administrative or legal mandates for the science curriculum and yet engenders in the minds of students the misconceptions promoted by antievolutionists. The most often heard and most damaging of these ideas are that there is evidence against the occurrence of evolution; that such evidence has given rise to a serious controversy in the scientific community over whether evolution happened; that evolutionary theory is, therefore, weak compared to other established scientific theories; that supernatural explanations can be part of science; and that such explanations, rather than evolution, should be used in modern biology.

Teachers may unintentionally promote some of these and other misconceptions if they cut back on teaching evolution due to pressure from students or from the community at large. I have spoken with public high school teachers who have yielded to such pressure, and most of them aren't proud of their actions and tend to remain quiet about them. When they do explain, they often say that they went into teaching to help kids get excited about science, not to get themselves and their students caught up in a culture war.

How does cutting back on teaching evolution cause misconceptions about evolution? First, students may notice that evolution is being slighted and may well conclude that there's not much to say on the topic. Second, and perhaps more important, many students are aware of the social and political issues surrounding the teaching of evolution and are likely to think that evolution is being neglected in school because the arguments against it are strong (that is, that evolution is a weak theory).

I have also talked with numerous public high school teachers who admit to being antievolutionists and who deliberately lead students to conclude that the theory of evolution is weak and controversial, either by directly telling that to students or by using pedagogical methods that result in students coming to these conclusions on their own. These teachers see evolutionary theory as a great evil and, overtly or covertly, bring an antievolution message into the public schools with missionary zeal.

Neither type of teacher is doing students a favor. The responsibility of public school biology teachers is to teach evolution thoroughly and correctly. The educational goal should be to have students acquire a scientific understanding that is compatible with the mainstream consensus of the scientific community. This means that students should come to understand that evolution is a central unifying principle in the science of biology; that the scientific community considers evolution to be a fact and not at all controversial; and that evolutionary science is as strong and well founded as any other science. Therefore, teachers should never disparage evolutionary science, despite any personal beliefs or any pressure from others. It is just as unsuitable for teachers to teach students that evolution is scientifically weak and controversial as it would be to teach that the Earth is only ten thousand years old, that humans and dinosaurs coexisted, or that scientific creationism or intelligent design creationism are valid science, because all of these notions have been resoundingly rejected by the leading scientific and science education associations.

Of course, teaching that evolution is a well-established science shouldn't prevent teachers from introducing students to genuine debates within the scientific community about the nature or relative importance of various mechanisms or patterns and processes for evolution. For example, some teachers might decide that discussing the debate about the pace and timing of evolution (punctuated equilibrium versus phyletic gradualism) is appropriate for their students.

The Controversy That Isn't, and Why Teachers Shouldn't Teach It

Students and the general public may have a host of misconceptions about evolution, but here I'll focus on the one that figures most prominently in the current debate about evolution in the high school curriculum. This is the misconception that there is significant scientific evidence against evolution and that, as a result, the scientific community considers the theory of evolution to be controversial.

Some antievolution groups, with the goal of strengthening this misconception among students, advocate that teachers "teach the controversy" (a controversy that, as far as the mainstream scientific community is concerned, doesn't exist). These groups frame the controversy not just as a debate about evolution, but as a debate between evolution and intelligent design. That is, they promote the idea that intelligent design is a legitimate scientific theory that seriously competes with evolution to explain biological facts.

Good science, of course, should always be open to alternative explanations, to different theories. Advocates of intelligent design take advantage of this by contending that high school science textbooks leave no room for discussion of alternative theories to evolution, so students should at least be made aware of the controversy and of intelligent design as an alternative theory. The first problem with this argument is that there is no scientific controversy about evolution, and the second problem is that intelligent design doesn't qualify as a scientific theory. Clearly, then, teaching students about a controversy concerning evolution and about intelligent design as an alternative theory to evolution would engender serious misconceptions.

Some antievolution advocates try to get around this difficulty by claiming that making students aware of a scientific controversy about evolution is not teaching. Rather, they say, it's like informing students of a bathroom closure or cancellation of gym class for the day. Teachers know, however, that this analogy doesn't make sense. The fact that a bathroom has been closed or that gym class has been canceled doesn't directly relate to anything in the academic curriculum. Making students aware of such things is irrelevant to academic learning. However, telling students that a particular scientific concept covered in their science class is controversial may directly affect their learning about that concept. This may not be *good* teaching, but it surely is teaching. (This view was strongly supported in the 2005 decision by Judge John E. Jones III in the case of *Kitzmiller v. Dover Area School District.*)

Science teachers are taught to determine students' misconceptions and then get students involved in activities that help them see their mistakes and lead them to construct more appropriate understandings. This approach to teaching science is based on accepted ideas about how students learn science, ideas that have been supported by the leading science and science education associations for decades. Both the National Science Teachers Association (NSTA) and the National Academy of Sciences (NAS) have noted that students bring misconceptions to the classroom about both scientific phenomena and scientific processes and that teachers need to engage these views in order to help students improve their understanding of science. And just as important, science teachers need to do their best not to *engender* misconceptions that students then have to confront at a later time. That's why students shouldn't be taught (or "made aware") that there is a controversy about whether evolution occurred or that intelligent design is a scientific alternative to evolution—teaching such falsehoods is poor pedagogy.

All the leading science education and scientific associations maintain that evolution is one of the most important concepts—if not *the* most important concept—in biology, and that students cannot attain a well-rounded background in science without learning about evolution. Telling students that intelligent design is a scientific alternative to evolution could give students the idea that evolution is weak, that it is inferior to and less well supported than other sciences. But this is far from true. There are no gaps, problems, or weaknesses that make the scientific community question the plausibility of evolution, and creating student misapprehensions about this would be poor science education indeed.

By now it should be clear that "teaching the controversy" is not about the concern for good pedagogy but about advancing the antievolution agenda. Rather than having students investigate true scientific controversies over patterns and processes within evolutionary theory, the goal of "teaching the controversy" is to have students think there is a genuine controversy within the scientific community about evolution.

Some antievolutionists take a different approach to promoting the idea that there is a scientific controversy about evolution. They maintain that teachers should lead students to question the occurrence of evolution because this encourages critical thinking. Naturally, most teachers want to encourage critical thinking, but that doesn't mean encouraging students to denigrate a whole science or deny well-established facts. Rather, encouraging critical thinking means leading students to question concepts for the purpose of improving their thinking skills in general. Also, having students engage in critical thinking helps teachers determine any underlying confusions that might relate to a particular learning goal. For example, a learning goal in math may be for students to understand why the mathematical community, given their norms (e.g., data, logic, procedures), has concluded that 2+2=4. To meet this goal, a teacher may have students critically analyze the math. If a student concludes that the norms of the mathematical community do not support 2+2=4, but rather 2+2=5, then the teacher facilitates learning activities to address the underlying false premise that led the student to this conclusion. The pedagogy involved focuses on appealing to the student's reasons and rationality, with a goal of enriching the student's sense of what the mathematical community considers good mathematical reasoning. Likewise, a learning goal for science students is to understand why the relevant scientific community, given their norms (e.g., data, logic, procedures), has accepted evolution as a fact. But the antievolutionists don't accept this consensus of the scientific community, and many are being disingenuous when they say they are trying to promote critical thinking. That is, their goal is not to help students understand why the scientific community accepts evolution as a fact but to have students question or deny evolution and to foster in students the misconceptions that intelligent design is a scientific alternative to evolution and that evolution is a weak science.

Should students critically analyze evolution in scientifically and pedagogically appropriate ways? Because of the strong possibility of students' religious objections to the concept of evolution, and because students' critical thinking skills can be improved in so many other areas of science, having students critically analyze evolution is ill advised and unnecessary. It is unnecessary because teachers can get around this difficulty by having students critically analyze concepts *within* evolutionary science. For example, students could analyze phyletic gradualism versus punctuated equilibrium; they could debate the question of which modern mammal group whales are most closely related to; or they could discuss the question of which reptile line gave rise to birds. Even though these questions presuppose that evolution has in fact occurred, students are likely to experience them as being less in conflict with their religious beliefs than the direct question of evolution.

Still another tack taken by antievolutionists is to suggest that teachers should let students know that evolution is "only a theory." Many people feel this is appropriate, but that opinion is based on a lack of understanding about the meaning of the word *theory*. In everyday conversation, *theory* means a hunch or a guess based on not much evidence, if any. In science, however, *theory* means something quite different. The National Academy of Sciences, for example, defines a scientific theory as "a well-substantiated explanation."[1] The leading scientific societies all accept that evolution is scientifically confirmed and that evolutionary theory is a well-substantiated explanation of a wide range of biological facts and processes, so it is clear that characterizing evolution as "only a theory" misleads students. Furthermore, intelligent design doesn't qualify as a scientific theory, since it isn't accepted as such by the mainstream scientific and science education communities.

Teachers under Pressure, and How They Should Respond

Many teachers are under pressure to infuse antievolution material into science curricula. This pressure may come from parents, from community groups, and even from administrative directives. In a recent poll by the NSTA, approximately a third of responding teachers indicated that they have felt this type of pressure.[2] However, individual teachers shouldn't—and most don't—unilaterally decide what concepts and theories are scientifically valid. Rather, they look to mainstream scientific associations for guidance on these questions. This is because most high school science teachers aren't scientists—they don't have research labs, don't publish in peer-reviewed scientific publications, don't present science at scientific research conferences, and don't receive funding for scientific research. This fact is particularly relevant when teachers hear of a supposed scientific alternative to something as important as evolution, a major unifying concept in biology.

As we have seen, when teachers examine the position of mainstream scientific organizations, they find that there is no scientific alternative to biological evolution and that intelligent design is not a valid scientific theory. The American Association for the Advancement of Science, for example, has a board resolution stating that "the lack of scientific warrant for so-called 'intelligent design theory' makes it improper to include as a part of science education."[3] Furthermore, in a recent NAS publication for science teachers, the NAS president states, "Opponents of evolution assert that the scientific justification for evolution is lacking, when in fact the occurrence of evolution is supported by overwhelming evidence. Legislators and school boards insert wording into laws, lesson plans, and textbooks mandating that evolution be taught as a controversial explanation of life's history, though no such characterization is scientifically warranted."[4] The consensus of the scientific community is clear to science teachers. Therefore, any policy requiring them to advance intelligent design education in science classrooms would necessarily also require them to disregard the consensus of the scientific community.

Likewise, teaching students about intelligent design requires teachers to disregard the recommendations of national professional science teachers' associations. For example, the National Association of Biology Teachers' (NABT) official statement on teaching evolution declares: "Explanations or ways of knowing that invoke non-naturalistic or supernatural events or beings, whether called 'creation science,' 'scientific creationism,' 'intelligent design theory,' 'young earth theory,' or similar designations, are outside the realm of science and not part of a valid science curriculum."[5] The official position statement of the NSTA on teaching evolution states: "Policy makers and administrators should not mandate policies requiring the teaching of 'creation science' or related concepts, such as so-called 'intelligent design,' 'abrupt appearance,' and 'arguments against evolution.' Administrators also should support teachers against pressure to promote nonscientific views or to diminish or eliminate the study of evolution."[6]

Not surprisingly, the use of an intelligent design textbook such as *Of Pandas and People* in the science classroom also goes against the recommendations of national teaching associations. The official NSTA "Background Paper on the Use and Adoption of Textbooks in Science Teaching" encourages "criteria that promote the use of textbooks that are... accurate in science content."[7] Since the major scientific organizations don't consider intelligent design to be valid science, it follows that any textbook that presents intelligent design as a valid science doesn't meet these NSTA criteria.

Also, teaching intelligent design goes against standards for public school teachers' professional preparation and professional development. For example, the Association for Science Teacher Education (ASTE) position statement on science teacher preparation and professional development states: "Science teacher preparation and professional development programs are essential elements in the success of contemporary science education.... These programs should focus on practices that are grounded in the research and professional literature on science learning and teaching."[8] And the NSTA position statement on science teacher preparation states: "NSTA strongly recommends that all science teacher preparation programs have a curriculum that includes substantive experiences that will enable prospective teachers to... understand how to find and use credible information... on the curriculum."[9]

Since intelligent design is not a credible alternative to evolution, it should be excluded from the science curriculum, and in fact, not a single mainstream science education organization has seen a need to consider how to teach intelligent design. For example, the program for a recent ASTE annual conference listed more than two hundred sessions, none of which included anything about teaching intelligent design. In contrast, of course, many sessions included presentations about teaching evolution. Likewise, at a recent NABT national conference, the American Institute of Biological Sciences and the Biological Sciences Curriculum Study held a major joint symposium on evolution education, with more than twenty-five scientists and science educators making presentations to an audience of science teachers. Many of the presenters discussed why intelligent design is *not* appropriate for inclusion in the science curriculum, while none contended that intelligent design is appropriate.[10] Finally, no national, regional, or occasional small conference of any of the leading science education associations has had sessions on teaching the nonoccurrence of evolution or on teaching the so-called science of intelligent design.

Many other official statements from the most prestigious science organizations and science education organizations discuss the teaching of evolution while also critiquing intelligent design and other forms of creationism. Again, nothing in any of those statements would suggest to a teacher that there is serious doubt about evolution. Nothing in these statements would lead teachers to believe they should be teaching intelligent design to their students. And all these associations hold that it is bad pedagogy to teach that evolution is only a theory or that there is evidence against the occurrence of evolution.

How, then, should teachers respond to pressures to include antievolution messages in their science classrooms? Given the clear status of intelligent design as a nonscience, it's equally clear that teaching intelligent design in secondary school science classrooms will improperly prepare students for postsecondary science education at secular schools—and therefore teachers shouldn't do it. High school students who are "made aware" that intelligent design is a scientific alternative to evolution will have to confront this misconception when they continue their science education in college. Certainly, students will find no science degree programs based on intelligent design at secular colleges or universities, or at good sectarian schools such as Baylor, Notre Dame, and Brigham Young. Furthermore, students who have been told that evolution is a theory in crisis (or not told about evolution at all) will be ill prepared for the way evolution is presented at the college level.

Preparing students properly for postsecondary science education includes helping them understand that evolution is as well supported and well accepted as any other scientific concept—for instance, the concept that the Earth goes around the sun or the concept that microscopic organisms can cause disease. In fact, teachers need to go beyond this—they need to have students actively engage the common misconception that evolution is only weakly supported. Good pedagogy in high school science education means helping students understand that there is no credible scientific evidence against evolution, that the mainstream scientific community considers evolution to be factual, and that evolutionary theory is well supported and well accepted.

Religion and the Teaching of Evolution

Ever since Darwin published *On the Origin of Species,* large numbers of people have perceived a conflict between religion and evolution, and this perceived conflict has affected secondary school science education in many ways. Legal questions, of course, come up whenever religious issues are involved in public education. But in the everyday life of the science classroom, two other questions come to the fore. First, how should teachers respond when students raise religious objections to evolution? And second, how can teachers be sensitive to some students' religious beliefs and moral values and still offer all students good science education?

The most notable recent legal case involving religion and evolution was the 2005 federal case of *Kitzmiller v. Dover Area School District,* in which a group of parents sued to prevent the Dover, Pennsylvania, school board from teaching students that intelligent design is an alternative scientific theory to evolution. In a landmark decision, intelligent design was deemed a religious theory, and thus *any* advancement of it was prohibited: "The Establishment Clause forbids not just 'teaching' religion, but any governmental action that endorses or has the primary purpose or effect of advancing religion."[11] In light of this decision, what should a public school teacher do if a student brings up intelligent design or other creationist ideas, either as an argument against evolution or to hear what the teacher thinks about it? In this situation, it would be pedagogically appropriate for the teacher to acknowledge that religious objections to evolution exist and to briefly explain why discussion of such objections isn't within the scope of the course. This wouldn't in any way advance religion, and it would avoid the pedagogical error of presenting religious objections as science.

Now consider something I've discussed previously: that some antievolution activists want teachers to use evolution as an example of a theory for which the evidence is weak. Of course, doing this would be poor pedagogy for *all* students because the evidence for evolution isn't weak, it's very strong. However, it would be *particularly* poor pedagogy to use evolution in this way for religious students whose beliefs conflict with evolution. Why? Because many of these students are likely to have strong negative feelings about evolution that will make it hard for them to be open to where the scientific evidence may lead. If the educational goal at hand is to help students understand how to evaluate scientific evidence and scientific theories, students will be best served if teachers avoid religiously charged examples like evolution. Singling out evolution to use in this way might advance the agenda of antievolution advocates, but it is also likely to inhibit the educational progress of many students.

Leaving aside the legal issue mentioned above, worse than singling out evolution as a weak science would be the direct inclusion in a science course of intelligent design or some other creationist science that students perceive to be God-friendly. (Many students are well aware that the Designer in intelligent design is assumed to be God.) Evolution, like every other legitimate science, takes no position concerning the existence or nature of an ultimate Designer. Including creationist material in the science curriculum creates or reinforces student misconceptions by suggesting that scientific explanations either support or detract from religious views in general and that evolutionary science in particular is incompatible with religion. This poor pedagogy is widespread in Christian education advocating intelligent design and other forms of creationism, but it should be absent from secular public school science education. Students will perceive that evolution doesn't posit a designer/creator while intelligent design explicitly does. The judge in *Kitzmiller* agreed, concluding that reasonable, objective students would view intelligent design as an endorsement of religion or a religious viewpoint. Intelligent design advocates have claimed that this conclusion is unwarranted, but consider how these advocates would react if evolutionary science were perceived to be God-friendly and intelligent design were perceived to be silent about God. How likely is it that they would still be clamoring for an alternative science?

The strength of students' emotional ties to their religious beliefs cannot be overemphasized, and neither can the potential for negative educational outcomes when students feel that their beliefs are in conflict with what they're being taught. It's well known that a segment of American society consists of Christian biblical literalists who believe that God, the heavenly Father, the Creator of the universe, loves them but that there are moral absolutes on which they will be judged by God when they die. If they pass God's judgment, they will enjoy eternal life in paradise after death, where they will be reunited with their loved ones who have died and passed God's judgment. Students with these beliefs who are taught (directly or indirectly) that evolution is against God or ignores God and that intelligent design or other forms of creationism embraces God or recognizes God are hardly likely to think clearly about evolution.

But even in schools that offer good evolution education, without any suggestion of a conflict between evolution and belief in God, teachers should be aware that many religious students are likely to bring to the classroom assumptions about the relationship between faith and science, especially evolution. Pedagogically speaking, one of the most widespread of these assumptions is that evolution is incompatible with the Bible. The vast majority of Christian students are not biblical scholars, as they themselves would generally be the first to admit. They claim to believe in the Bible, but they know very little about its contents—to the disappointment of their clergy. If their church rejects evolution, students are unlikely to know the precise biblical reasons why. Often, students don't even know exactly where they heard that there is biblical justification for denying evolution—maybe from church leaders, parents, and friends, or maybe from more specialized sources such as creationist publications. Whatever the source, the basic message is the same—evolution is against what is in the Bible, and the Bible comes from God.

Science instructors who don't share this assumption—that evolution and the Bible are in conflict—may have trouble understanding why students have such strong feelings on the subject and why many students feel a strong moral obligation to fight against the teaching of evolution. Teachers can get some idea about this issue by imagining how they might feel if students were being taught something clearly false and reprehensible—for example, that one race is inferior to another, that the Holocaust never happened, or that people with AIDS deserve to suffer and die. All the teachers I know would combat such outlandish teachings as forcefully as they could, and this is precisely the point—many students with religious beliefs like those described above feel that evolution, particularly what they consider to be Creator-less evolution, is just as false and reprehensible as racism and just as wrong to teach. The way for science instructors to deal with this issue is by helping students understand that evolutionary science doesn't deny the existence of a supreme being—that evolution simply doesn't address such nonscientific questions. This should be done, however, without directly introducing intelligent design, with its Creator or Designer language, into science lessons. Bringing the intelligent design concept and language into the classroom will unavoidably give students the impression that questions and answers about a Creator or Designer *are* part of evolution instruction—and that evolution doesn't mention a Creator while intelligent design does. It would be poor pedagogy in the science classroom to needlessly arouse students' strong feelings about religion.

Some intelligent design advocates contend the opposite—that bringing these emotions into the classroom will promote greater interest among students and result in greater learning. While it may be true that student interest will increase (feeling the need to defend one's religious beliefs often increases interest), it is not true that student understanding of science will also increase. Instead, students will become confused about how science works; that is, they will have previously been taught that methodological naturalism is the underpinning of modern science, but the inclusion of intelligent design in their coursework will imply that supernaturalism can supply plausible scientific explanations.[12]

Some intelligent design advocates believe they have a compromise position: Science teachers should tell students that intelligent design is a scientifically appropriate alternative to evolution but shouldn't then answer students' questions about intelligent design. It should be apparent how pedagogically irresponsible this position is. No public school science teacher should be put in the position of telling students, in class, about the existence of a scientifically valid theory that supposedly counters a cornerstone of biology and then not be able to answer students' questions about such an obviously important scientific theory. As the judge in *Kitzmiller* made clear, a reasonable student would conclude that intelligent design is a kind of secret science. This illogical practice should not exist anywhere in public school science education.

Some people may feel that it is appropriate to compare intelligent design to evolution in the science classroom because certain educators have suggested that it is good for students to directly compare their misconceptions with accurate science. There are many problems with this form of pedagogy as it relates to evolution education for students of high school age. The most disturbing is that teachers would be intentionally comparing science with students' religious beliefs, which would arouse students' emotions in educationally counterproductive ways. Large numbers of students have been taught about the goodness and theological correctness of intelligent design (or other forms of creationism) in their places of worship, and most students previously unaware of intelligent design will immediately recognize it as bringing their religious beliefs into their science lessons. It's true that mainstream science curricula sometimes include coverage of William Paley's 1802 model of intelligent design (which differs from the model used by the modern intelligent design movement); however, Paley's model is included *only* to increase student understanding of why evolution is accepted according to the norms of scientific disciplines and Paley's model is not.

If the objective is to have students confront common misconceptions in science, then teachers can accomplish this in other ways—for example, by having students analyze the question of whether humans and dinosaurs coexisted or not. This is proper even if students acquired the false belief through religious training. Of course, confronting misconceptions should be done in a way that doesn't advance a particular religious view and that isn't directly or explicitly connected to one (for example, intelligent design). Dealing with a particular religious view wholesale, however, is inappropriate—any good that could come from directly comparing modern intelligent design to evolution would be outweighed by the pedagogical harm to students.

Various intelligent design teaching methods are often promoted as good pedagogy within fundamentalist Christian schools and at their school conferences. The largest organization of Christian schools is the Association of Christian Schools International, which has about fifty thousand teacher members, includes more than five thousand schools and 1.2 million students, and holds more than twenty-five annual conventions. At such conventions, private school science teachers can obtain information on teaching about intelligent design, can hear the so-called evidence against evolution and about "the controversy," and so forth. In public schools, the philosophy of education is neither to endorse nor to detract from any religion, but Christian schools generally have an explicit philosophy of Christian education. Often, this philosophy includes integrating theological beliefs with the full range of knowledge (including scientific knowledge) and with the methods of acquiring knowledge (including scientific methods). In keeping with this philosophy of education, many Christian schools use biology textbooks different from those used in public schools. These texts include the aforementioned *Of Pandas and People*, the seven-hundred-page *Biology for Christian Schools*, published by Bob Jones University Press, and the four-hundred-plus-page *Biology: A Search for Order in Complexity*, published by Christian Liberty Press. All of these textbooks claim that the natural world is the product of divine creation or intelligent design. Teacher ancillaries are available from various sources to help teachers facilitate student learning about why accepting the fact of evolution is a bad idea. This indicates how teacher professional preparation and development can differ, depending upon whether the educational goals for the science classroom are religious or nonreligious.

Public school science teachers encounter interesting problems when students come from private or home schools, whether Christian or not, where the concept of evolution is considered false or exceptionally suspect scientifically. Students may have learned about evolution in their previous schooling, but the primary point taught was that evolution is scientifically lacking compared to intelligent design or other forms of creationism. Their science teachers focused on having these students better understand that intelligent design is at least a valid scientific alternative to evolution, if not far superior. In line with this objective, these students were probably taught one or more of the following ideas in their science courses: that organisms are too complex and orderly to have come about by way of natural processes; that various forms of life appeared abruptly through an intelligent agency—fish suddenly appeared with fins and scales already intact, and birds suddenly appeared with feathers, beaks, and wings; that scientists have come to realize that evolutionary explanations are insufficient and are now seriously debating intelligent design in the scientific community; that good, open-minded scientists realize that natural explanations are insufficient to explain the diversity of life, in contrast to close-minded scientists with a materialistic philosophy, who hold on to evolutionary theories; and that the evidence against evolution is significant and has led the scientific community to question its scientific factuality.

Students with misconceptions such as these bring unique challenges to public school science instructors.[13] Not only were these students taught arguments against evolution directly and purposefully by people they probably respect, but this teaching may receive ongoing parental and religious organizational support. Unlike other student misconceptions, these weren't learned through misunderstanding, but learned as correct information from previous science teachers and science materials.

Teachers and the Battle of Creationism Versus Evolution

Many science teachers see the creationism versus evolution controversy as analogous to an academic dispute, and this perception can make it difficult for them to fully appreciate its effects in the classroom. Even the most intense and emotionally charged academic controversy doesn't come close to arousing the strong emotions and seriousness of resolve felt by many creationist leaders, lay creationists, and creationist students concerning the teaching of evolution. Generally, science teachers don't view creationist arguments as academically on par with arguments for evolution—usually, quite the opposite. But thinking that they can quickly resolve a creationism versus evolution controversy in a classroom by using some standard academic method is a bad mistake based on a serious underestimation of the nature and strength of the differences between camps.

School board and court battles may continue, but the reality of what happens when the classroom door closes is paramount. Intelligent design advocates appear to understand this, as they have been attempting through the students themselves to influence what goes on in the classroom. Anyone who doubts that powerful, professional antievolutionists are interested in creating a more student-centered front should read the works of Phillip Johnson, a Berkeley law professor and the father of the intelligent design movement. In his very popular antievolution book *Defeating Darwinism by Opening Minds*, Johnson states that his desired audience is high school students and college undergraduates. And in his first antievolution book, *Darwin on Trial* (on which *Defeating Darwinism by Opening Minds* is based), Johnson characterizes evolution education as a campaign of indoctrination in the public schools and emphasizes that something should be done about it. I have looked in hundreds of secular bookstores across North America, including both college bookstores and popular bookstores, and this antievolution book has always been shelved not in the religion section of the store but in the science section—typically among the evolution books!

Many science instructors think that there is no reason for them to be concerned about creationists and their ideas, but they are wrong. They underestimate the fierce emotions involved in the creationism versus evolution issue, and they fail to realize that a war is being waged against evolution in our classrooms. In addition, science instructors often underestimate creationists, believing they are a small, disorganized, and minimally funded group with a minuscule audience, but the contrary is true: antievolutionists are well organized, well funded, and numerous enough to have already caused significant harm to the integrity of science education. Also, they have large audiences, and, of most direct importance to science instructors, they appear to believe they are in a battle with those who teach evolution.

As science teachers come to understand these realities, they will need to decide how to respond. With the religious controversy over the teaching of evolution reported in the media, with parents confronting their children's science teachers on this issue, and with students themselves confronting their instructors in high schools and colleges, would it be best—and easiest—to just stop teaching evolution in the classroom? Can't students attain a well-rounded background in science without learning about this controversial topic? The overwhelming consensus of biologists in the scientific community is no. Students need to learn about evolution because it provides the context and unifying theory that underpins and permeates the biological sciences. Without instruction on evolution, students would learn disparate facts but not learn about the thread that ties them together, and they would not be helped to answer underlying scientific or proximate *why* questions. Without an understanding of evolution, they won't understand processes based on evolution, such as insect resistance to pesticides and microbial resistance to antibiotics; they won't come to understand how evolution connects to other scientific fields; and, ultimately, they won't fully understand the world of which we are a part. Understanding evolution is one of the most important aspects of attaining scientific literacy.

It's vital that teachers understand that teaching of evolution should not be diminished but in most cases increased. It's also vital for teachers to understand that this can be done while respecting student religious convictions—it takes good teachers with good training.

Teaching and Learning in the Science Classroom

Students come to the science classroom with their own particular memories, knowledge, experiences, and conceptions about evolution, regardless of whether they have taken courses in biology or general science. Often, some of their conceptions—about evolution and other science topics—are scientifically inaccurate, developed on the basis of what they've read in the popular press, found on the Web, seen on television, or gathered from their interactions with nature, their peers, and their parents and other authority figures. Instructors at all levels have a complex task: to tap into what students know in order to assist them in building on their scientific knowledge, to help them replace their scientific misconceptions with scientifically useful conceptions, and to help them construct meaning from their learning experiences.[14]

This last task—helping students construct meaning from their learning experiences—is a daunting one. Only the students can construct meaning for themselves; instructors cannot fill them with knowledge. All instructors can do is to provide learning experiences that help students generate links between relevant information they already know and new information. From such a *constructivist* perspective, learning is a social process in which teachers help students make sense of experience in terms of what they already know. But how does a teacher engineer lessons to help students link the new information with the old in useful, accurate, and appropriate ways? And what happens if the old information consists of one or more scientific misconceptions?

Many instructors know from their own classroom experience that students' convictions are not easily changed. The process of conceptual change takes a long time, perhaps years, depending on the concept, and it appears to be an incremental process. In learning evolutionary concepts in particular, students seem to need an extended exposure to and interaction with these concepts for growth of their understanding to occur. Each learning experience that instructors provide for students is only one steppingstone on the path toward a more complete understanding.

As part of the constructivist approach in the classroom, instructors should provide situations in which students examine the adequacy of their prior scientific conceptions by arguing about and testing them. During this process of arguing and testing, the students may come up against contradictions that provide the opportunity for them to acquire more appropriate concepts. As students practice this process, they also become increasingly skilled in the procedures used in concept acquisition.

When teaching evolution, then, instructors must first uncover and understand students' scientifically naive conceptions about evolution, which they can do by analyzing students' answers to evolution problems or questions. Next, instructors need to provide students with learning situations that help them examine their mistaken ideas and, ultimately, become dissatisfied with them. These situations might involve focusing students' attention on observations, data, or problems that lead students to realize that their existing conceptions do not explain the observations or data or are not useful in solving the problems. During this process, instructors must help students think through their dissatisfaction and realize why their naive conceptions are not working. Student-to-student and instructor-to-student discussions can be important in this regard, and teachers should engineer such interaction to challenge students to think and talk about their scientific ideas, while guiding students to see the relationships and contrast between their own scientific ideas and the ideas widely accepted by scientists, which may well be new ideas to students. Finally, instructors should help students examine these new ideas by focusing students' attention once again on observations, data, or problems to see if these ideas are useful in the explanatory or problem-solving contexts in which their prior conceptions were not helpful.[15] Various teaching methods appear beneficial in accomplishing these goals, and teachers should take every opportunity to learn about, evaluate, and experiment with them for themselves, to see if they fit the needs of their students. As mentioned previously, having high school students compare a specific scientific misconception (for example, dinosaur/human coexistence) with the correct science concept is appropriate, but comparing a religious theory (for example, modern intelligent design) with a scientific theory is inappropriate.

Whatever methods teachers use to teach evolution, they should remember that students differ greatly in their abilities to process information—that is, some students process information most readily in certain ways, while other students do so in different ways. Another way of talking about this is to say that people have different types of intelligences to different degrees. Howard Gardner sketched this idea in his 1983 book *Frames of Mind: The Theory of Multiple Intelligences,* in which he outlined a theory of human intellectual competencies that has continued to develop over two decades and is now an important educational idea for the twenty-first century. The theory of multiple intelligences holds that intelligence is not just a single property of the human mind; it is a substantial set of largely independent attributes. This concept has been praised by scientists as eminent as the late Stephen Jay Gould.[16] In 1999, Gardner offered this definition of an intelligence: "a biopsychological potential to process information that can be activated in a cultural setting to solve problems or create products that are of value in a culture."[17] The nine intelligences are linguistic, logical-mathematical, musical, bodily-kinesthetic, spatial, interpersonal, intrapersonal, naturalist, and existential.

In the context of the multiple-intelligences theory, Gardner suggests that instructors engage students using entry points that align roughly with the intelligences. With regard to evolution, useful types of entry points include:

- NARRATIONAL—*engages students who enjoy learning through stories. For example, develop lessons that use the story of Darwin's voyages or the life course of a particular species.*

- QUANTITATIVE/NUMERICAL—*engages students who are intrigued by numbers. For example, develop lessons that look at numbers of individuals of a species in varied ecological niches and at how those numbers change over time.*

- LOGICAL—*engages students who think deductively. For example, develop lessons based on if/then syllogisms that support the concept of natural selection.*

- HANDS-ON—*engages students who enjoy manipulating materials and carrying out experiments. For example, have students breed generations of* Drosophila *(fruit flies) and observe the incidence and effects of genetic mutations.*

- SOCIAL—*engages students who learn best by interacting in groups. For example, give a group of students a problem to solve, such as determining what happens to various species in a given environment following a dramatic change in climate. Role-play also engages this type of student.*

Summary and Concluding Thoughts

No state or national science education standards, benchmarks, or frameworks support teaching (a) intelligent design; (b) that there are gaps or problems that cast doubt on the scientific consensus about evolution; (c) that a controversy has the scientific community seriously debating the fact of evolution; (d) that organisms are too complex and orderly to have come about by natural processes; (e) that various forms of life appear or have appeared abruptly through an intelligent agency; or (f) that supernatural explanations are part of modern science. Furthermore, as was decided in the *Kitzmiller* case, "making students aware" in a science course that intelligent design is a scientific alternative to evolution or that a serious scientific debate exists about whether evolution occurred and then effectively muzzling teachers so they may not answer student questions is pedagogically irresponsible. Also, singling out evolution as a science supported by weak evidence sends an inaccurate signal to students that evolution is an inferior science.

William Paley's 1802 model of intelligent design has been used in advocated pedagogy, but the goal of those lessons is to have students learn why evolution is an accepted science and Paley's intelligent design is not. Public high school science educators should not imply that good, open-minded scientists accept intelligent design or other forms of creationism because they realize that natural explanations are insufficient to explain the diversity of life, while bad scientists reject intelligent design or other forms of creationism because their minds are closed to supernatural explanations. Whether deliberately encouraged or not, these kinds of misconceptions are detrimental on multiple levels. Overall, pedagogical approaches that advance intelligent design or other forms of creationism or that diminish, denigrate, or disparage evolutionary sciences are not accepted by the relevant science education and scientific communities.

At the beginning of this chapter, I noted that helpful court case decisions, school board policies, administrative directives, and positions of leading science education associations are naturally all very important for evolution education. However, as educators know well, when the classroom door closes, it is the individual teacher who is really in control of that learning milieu. It takes good teachers with the right mix of abilities, education, dedication, and sometimes courage to facilitate students' understanding of evolution. These champions make the biggest difference in young people's education. In contrast, teachers may do a barely adequate or even a poor job of teaching evolution, and student learning will suffer greatly if so.

Given the increased antievolution pressures dedicated teachers are feeling from various segments of the public, school administration, politicians, and the media, more teachers may ultimately cut back on their teaching of evolution. Whether taught directly or indirectly, students may learn misconceptions such as the ones I just listed.

Dubious teaching might also increase for other reasons. I believe that the greatest potential threat to evolution instruction comes from those who become public school biology teachers with the intention of diminishing evolution education. If large antievolution organizations are able to rally enough young people around this cause, we could conceivably have a situation where the majority of high school biology courses are taught by instructors who are against evolution education—with devastating consequences. This outcome is, thankfully, unlikely; however, there are already large numbers of public school biology instructors who teach for precisely this reason, even if they remain faithfully quiet about their mission.

Evolution education is terribly misunderstood and widely attacked. Good evolution education is the solution to this important problem. Evolution needs to be taught more in schools, not less. Current and future teachers need good evolution education and good pedagogical training on how to better address students' misconceptions about evolutionary science. Administrators, parents, and the general public need to stand up and actively support—cheer on—high school biology teachers for teaching evolution thoroughly and skillfully. After all, when that classroom door closes, the teachers are the ones who truly make the difference.

6: Defending the Teaching of Evolution: Strategies and Tactics for Activists

GLENN BRANCH

The Challenge of Defending Evolution

Eighty-one years after the Scopes trial, the media is filled with stories about battles over the place of evolution in the public school science curriculum. Unfortunately, not all the stories are mere nostalgic descriptions of the eight scorching days in July 1925 when John Thomas Scopes was on trial in a Tennessee courtroom for violating the state's Butler Act, which forbade teachers in the public schools "to teach any theory that denies the story of the Divine Creation of man as taught in the Bible, and to teach instead that man has descended from a lower order of animals." The issues involved in the Scopes trial are, alas, alive and well in the United States of the twenty-first century. Indeed, the Scopes trial is usually cited in the media as a harbinger of the current battles over evolution, whether in the classrooms, the courtrooms, or the legislatures. There is, unfortunately, no shortage of such battles on which to report.

Three prominent episodes have captured the headlines recently. In Dover, Pennsylvania, in October 2004, the school board adopted a policy providing that "students will be made aware of gaps/problems in Darwin's Theory and of other theories of evolution including, but not limited to, intelligent design"; the policy was ruled unconstitutional by a federal court in December 2005, and a newly constituted school board decided not to appeal the decision.

The board of education in Cobb County, Georgia, voted in 2002 to have labels affixed to textbooks that warned students, "Evolution is a theory, not a fact, regarding the origin of living things"; the policy was ruled unconstitutional by a federal court in January 2005, but the board of education chose to appeal the decision, so the ultimate fate of the disclaimer is not yet decided as of this writing. And in Kansas, in November 2005, a creationist majority on the state board of education, under the tutelage of a local intelligent design organization, engineered the adoption of a set of state science standards in which the scientific standing of evolution is systematically impugned. It remains to be seen whether voters in Kansas will eject the board members, as they did after a similar incident in 1999.

Dover, Cobb County, and Kansas were only the tip of the iceberg. In 2005 alone, the National Center for Science Education recorded more than eighty controversies in thirty states over evolution, primarily concerning the place of evolution—and supposed alternatives to it, such as creation science and intelligent design—in the public school science curriculum. Part of the reason for the persistence of the problem is the existence of a massive, if loose-knit, network of creationist groups that produces a steady stream of antievolution propaganda and that inspires, and occasionally coordinates, antievolution activity around the country. But the movement prospers because of the prevalence of creationist sentiment. According to a pair of recent national polls, a majority of people—60–65 percent—favors teaching creationism along with evolution, while a large minority—37–40 percent—favors teaching creationism instead of evolution.[1]

In light of such broad support, it is no surprise that creationism is politically connected—in August 2005, replying to a reporter's question about teaching intelligent design in the public schools, President George W. Bush said, "Both sides ought to be properly taught . . . so people can understand what the debate is about," and Senators Bill Frist and John McCain subsequently echoed him. Creationism is financially healthy—the revenues of the leading creationist organizations are in the seven figures, and there are hundreds of smaller organizations and thousands of fundamentalist churches that are sympathetic to the cause. And above all, creationism is able to enlist a dedicated cadre of proponents. Regarding evolution as a threat to the fabric of society, to their religious faith, and even to their hopes for salvation, creationists are able to exploit fervent, profound, and sincere religious beliefs, mobilize resentment and suspicion of the scientific and educational establishments, and muster their force where it matters most: at the grassroots.

With all of those advantages, what is there to stop the encroachments of creationism in whatever form on the integrity of science education? The lack of scientific worth of its supposed positive contributions to science and the lack of scientific merit of its supposed critiques of evolution (see chapter 2), to be sure. The fact that creationism is not uniformly accepted by people of faith (see chapter 3), with a number of prominent individual clergy and mainstream denominations actively opposing it. The weight of the case law holding that teaching creationism in the public schools violates the First Amendment's Establishment Clause (see chapter 4). The consensus of the educational community that science education is compromised when creationism is taught or evolution is downplayed (see chapter 5). But most of all, there is you. It is only with the help of dedicated and informed activists at the grassroots level that the battle to defend the teaching of evolution in the public schools can be fought. Along with the rest of *Not in Our Classrooms*, the present chapter is intended to help you join the fight.

The Battlegrounds

Any public exposition of evolution[2] is likely to elicit a backlash. For example, a decision to add a creationist display in the Tulsa, Oklahoma, zoo was recently made and then reversed; a handful of theaters associated with museums have reportedly declined to screen several IMAX films due to their evolutionary content; and there seems to be a growing trend in which creationists conduct their own tours of natural history museums in order to inoculate fundamentalist visitors against the evidence for evolution on display.

When the exposition is national and extensive, as it was for the PBS series *Evolution*, broadcast in 2001, the backlash is tremendous: two of the largest creationist organizations (the young-Earth creationist ministry Answers in Genesis and the Discovery Institute, the headquarters of intelligent design) attacked the episodes as they aired and subsequently published books devoted to criticizing the series. Examples could be multiplied. Nevertheless, today, as in the Scopes trial more than eighty years ago, the controversy over evolution is primarily focused on the public school science curriculum.

Creationist attempts to undermine the teaching of evolution occur at every level of public education, from the individual classroom through the local school district to the state government, or even, rarely, the federal government itself.[3] In the individual classroom, teachers may themselves be creationists, or may mistakenly think it fair to present creationism along with evolution, or may decide to omit evolution to avoid controversy. In a 1999 survey of Oklahoma biology teachers, for example, 12 percent favored teaching creationism only, 42 percent favored teaching creationism and evolution, and 22 percent favored teaching neither. And in a recent informal survey among members of the National Science Teachers Association, 30 percent indicated that they experienced pressure to omit or downplay evolution and related topics from their science curriculum, while 31 percent indicated that they felt pressure to include nonscientific alternatives to evolution in their science classroom.[4] Such pressure can come from students, parents, administrators, or members of the board of education. Whatever the source, if nobody is paying attention, poor evolution education in individual classrooms can go unchecked. In Bristol, Virginia, for example, a teacher recently agreed to stop using a self-published textbook entitled *Creationism Battles Evolution*—after using it for fifteen years without recorded protest.

Because the local press often reports on school board meetings, antievolution proposals adopted by administrators or board members are less likely to escape notice. Two of the three incidents mentioned above—Dover, Pennsylvania, and Cobb County, Georgia—involved actions taken by school boards that compromised the teaching of evolution in their districts in response to the antievolution sensibilities of a segment of their constituencies, even over the protests of their own science teachers. Owing to the increased likelihood of media scrutiny of such actions, it is common for such proposals to undergo hasty refinement. As introduced, they may call for teaching creation science or intelligent design, but after their proponents realize that such policies are constitutionally problematic, they often rewrite them to require disclaimers about evolution (as in Cobb County) or call for teaching the controversy about or critical analysis of evolution—that is, teaching evolution in such a way as to instill scientifically unwarranted doubts about it.[5] In states where local districts decide what textbooks to use, biology textbooks are sometimes challenged: in Dover, it was a board member's complaint that a standard textbook was "laced with Darwinism" that began the controversy.

At the state level, the most conspicuous activity is often legislation. During 2005, antievolution legislation was introduced in at least a dozen states, including calls for the teaching of creation science, the teaching of intelligent design, or forms of teaching the controversy about evolution. Such bills often are introduced as a sop to a creationist constituency and usually languish and die in committee, but their defeat is by no means guaranteed. In those states that have a centralized textbook adoption process, creationists often challenge the proposed biology textbooks for their evolutionary content. Such challenges are routine in Texas, where, for example, a husband-and-wife team successfully undermined the treatment of evolution in textbooks during the 1970s and 1980s. Since the state is the second-largest textbook market in the country (after California), any weakening in the treatment of evolution in the Texas textbooks to mollify Texan creationists is likely to appear elsewhere as well. Fortunately, in the last bout of biology textbook adoption in Texas, in 2003, all eleven proposed textbooks were approved despite a protracted creationist campaign.

A relatively new arena at the state level is state science standards, which provide guidelines for local school districts to follow in their individual science curricula. In the 1990s, as states began to introduce state science standards, the treatment of evolution was improving, penetrating even to districts where creationism was taught—Supreme Court or no Supreme Court—or where evolution was downplayed or omitted altogether. The importance of state science standards was cemented by the federal No Child Left Behind Act, passed in 2001, which requires states to develop and periodically revise standards. Because the content of statewide tests is based on the state standards, and because local school districts have a strong incentive for their students to perform well on such tests, the state standards matter. Moreover, if evolution is included in the standards, teachers faced with pressure to downplay or omit evolution are able to cite the standards by way of justification and authorization. In addition to Kansas, where creationists were successful in tampering with the standards in 1999 and again in 2005, Alaska, Georgia, Minnesota, New Mexico, Ohio, South Carolina, and West Virginia have also recently experienced controversies about the treatment of evolution in state science standards.[6]

As described in chapter 1, creationism continues to evolve, but the ancestral species still coexists with its descendants. Creation science is still taught in individual classrooms, and proposals to require equal time for creation science occasionally are introduced at higher levels—in the Mississippi state senate in 2005, for example. Proposals to require or allow the teaching of intelligent design are not uncommon, although the decision in *Kitzmiller v. Dover* is likely to deter a number of local boards of education from enacting such proposals. The fallback strategy, as always, is to impugn evolution while remaining silent about the supposed creationist alternative. Ways of implementing the strategy include requiring evolution disclaimers (whether written as in Cobb County or oral as in Dover), assailing the treatment of evolution in textbooks and state science standards (as in Texas and Kansas, respectively), and calling for critical analysis or objectivity in the teaching of evolution. It is the fallback strategy that is likely to dominate the creationism/evolution controversy in the future, but activity aimed at promoting creation science and intelligent design is bound to continue.

In whatever venue, three rhetorical themes are in constant use by creationists. During the Scopes trial in 1925, William Jennings Bryan contended that evolution is unsupported by, or actually in conflict with, the facts of science; that evolution is intrinsically antireligious, and in particular anti-Christian; and that it is only fair to take the desires of the taxpayers into account while developing the science curriculum. These three pillars of creationism, as they have been dubbed, have formed a sturdy platform for antievolutionism from Bryan's time to ours. Claims that evolution is a theory in crisis, that evolution is atheistic, anti-Christian, and immoral, and that it is only fair to teach both sides invariably appear in any arena in which the creationist movement is active. Keeping the pillars of creationism in mind will help you to understand—and to rebut—the bulk of the creationist arguments you are likely to encounter.

Eager to portray evolution as teetering on the verge of scientific collapse, creationists have constantly argued that evolution is a theory in crisis. One common way of doing so is to describe evolution as a "theory, not a fact" (as in the Cobb County disclaimer), in order to suggest that it is mere speculation or conjecture. Creationists also frequently allege that evolution is incompatible with established results from other branches of science or incapable of answering as-yet-unanswered questions about the history of life. Creationists are also fond of arguing inappropriately from authority, amassing a list of scientists who reject or doubt evolution or by quoting scientists out of context in order to suggest that they, too, reject or doubt evolution. It is not feasible to list and rebut all such claims here, of course, although there are resources, in print and on the Web, that attempt to do so.[7] It is in any case a valid response to note that the scientific community disagrees with the allegation that evolution is scientifically unfounded: the National Academy of Sciences—the country's most prestigious scientific organization—describes evolution as "thoroughly tested and confirmed . . . strong and well-supported . . . one of the strongest and most useful scientific theories we have."[8]

For the most part, the opposition to evolution is religiously motivated.[9] Unsurprisingly, then, creationists often allege that evolution is intrinsically atheistic, or that evolution is intrinsically anti-Christian, or that evolution is fraught with immoral consequences. But the fact that biology no longer invokes God to explain the history of life is no more proof of atheism than is the fact that meteorology no longer invokes God to explain the course of a hurricane. Moreover, many Christians—including prominent religious leaders such as the late pope John Paul II—and a number of Christian denominations instead regard their acceptance of evolution as compatible with, or even enriching, their religious faith.[10] Recently, more than ten thousand members of the Christian clergy signed a statement affirming the compatibility of evolution and Christian faith.[11] Accepting evolution is not logically equivalent to rejecting morality, and in general there is no plausible causal connection between accepting evolution and its supposed immoral consequences; the creationist quarrel is really with modernity and the social disruptions that followed in its wake.

Americans believe in fairness, and the most rhetorically effective of the three pillars of creationism is the appeal to fairness. If there are two sides to every issue, creationists argue, then why not present the other side of evolution? Creationists who have despaired of having creation science or intelligent design taught in the public schools still appeal to fairness, calling for teaching "the strengths and the weaknesses of evolution," or "the evidence for and against evolution," or, simply, "the controversy."[12] The ulterior hope behind the fallback strategy is that if evolution is presented to students as a theory in crisis, they will retain (or acquire) a belief in creationism. The power of the appeal to fairness is so strong that it is wisest to reply in kind: there is nothing fair about the creationist ambition for public education. It is not fair to citizens of a republic in which a basic constitutional principle is the government's religious neutrality. It is not fair to taxpayers, who run the risk of footing the legal bills due to a lawsuit over actions that compromise the teaching of evolution. It is not fair to teachers, who have a professional duty to teach in accordance with the scientific consensus. Most important, it is not fair to students, whose scientific literacy is on the line.

Taking Action

How to take action? Whether the problem is in the individual classroom, at the local district level, or at the state level, a few general principles apply.

- THINK LOCAL. *Just as all politics is local, so is all activism. It is you, and people like you—local citizens, acting locally—who will have the most impact locally.*

- DON'T GO IT ALONE. *If you are the only activist, there is a risk that policymakers will regard you as a lone crank. Although it is possible to do it all yourself, especially if the problem is at the individual classroom level, the proverb is true: in union there is strength.*[13]

- UNDERSTAND THE SYSTEM. *Ascertain the responsibilities, policies, and timetable of the school, school district, legislature, board of education, or department of education, as appropriate. You need to know what is happening, when it is happening, who is making it happen, and how you can arrange to have a say.*

- BECOME A PACKRAT. *Any piece of information about the policymakers or the creationist opposition may prove to be useful. Take notes at, or even record, relevant meetings and events; collect and file material that appears in print and on the Web.*

- BE CIVIL, DIPLOMATIC, AND FRIENDLY. *In dealing with policymakers, you want to be perceived as a friendly adviser, not a hostile critic. Be calm and polite rather than outraged; inquisitive and informative (but still concerned) rather than combative; helpful and cooperative (a useful phrase is "What can we do . . .") rather than confrontational.*

- BE INFORMED ABOUT THE ISSUES. *Having a copy of* Not in Our Classrooms *is of course a good start!*

ORGANIZING

A group of allies is useful for problems at the level of the individual classroom; for problems at any higher level, it is essential. Recruit members where you can find them: not only among your family, friends, and colleagues, but also among the natural constituencies for evolution education—science teachers and their organizations; professors and students at the local colleges and universities; local businesses with a scientific bent; the state academy of science and its members; natural history museums, zoos, aquaria, and science centers, and their staff; theologically liberal and moderate churches and synagogues; atheist, skeptic, and humanist groups; local chapters of civil liberties and church/state separation groups. Scientists, educators, educational administrators, members of the clergy, journalists and media professionals, and lawyers are all especially valuable allies. Choose a name for the group that reflects its goal: "—— Citizens for Science" is a popular formula, following the model of Kansas Citizens for Science, which helped to reverse the fiasco involving the Kansas science standards in 1999. The aim is to assemble a coalition that is focused on the basic goal of defending evolution education by convincing policymakers not to accede to creationist proposals, or—failing that—by supporting electoral campaigns against those policymakers (if elected) or lawsuits challenging the constitutionality of those proposals (if enacted).

In assembling your group, assistance from national or state organizations is useful, although it is no substitute for the work of grassroots activists. National organizations are often willing to help you get the word out to their local members, and they often have resources, in print or on the Web, that can be used. Again, seek help from the natural sources. Organizations that specialize in defending the teaching of evolution in the public schools are, of course, going to be especially helpful. There are such state-level groups now in Alabama, Colorado, the District of Columbia, Florida, Georgia, Iowa, Kansas, Maryland, Michigan, Minnesota, Nebraska, New Mexico, Ohio, Oklahoma, Pennsylvania, and Texas. These state-level groups vary in size, clout, and degree of activity, but they typically include at least a handful of experienced and dedicated activists, so they are always worth consulting. The only such national organization is the National Center for Science Education, which was founded in 1981 and is still the leading organization defending the teaching of evolution.

Your group can be as organized or as unstructured as necessary, but the essential functions are threefold: strategy, membership, and communications. The strategy committee is, as its name suggests, responsible for developing specific ways for the group to pursue its goal. The membership committee is responsible for expanding the membership base and for maintaining communications within it —telephone trees are traditional, but in the days of the Internet, e-mail and listserves are increasingly important, so make sure that the membership committee includes a technically savvy member and try to ensure that all of your members have e-mail and access to the Web. The communications committee is responsible for the group's communications with the public—face-to-face, in print, and on the Web—and with the media. (See Spreading the Word, below.) Obviously, it is helpful for the communications committee to include both technically savvy members and members with media experience. In general, your group is likely to have a wide diversity of skills, talents, interests, and levels of commitment. Try to find the right person for the right job: activists who know that their contributions are valuable and valued are going to be more contented, more committed, and more effective.

ORGANIZING A PERMANENT GROUP

A great deal of important work in defense of evolution education is accomplished by ad hoc groups operating mainly with resources donated by the activists themselves. If the controversy is persistent or recurrent, however, you may feel the need to incorporate as a nonprofit group. Although it imposes a certain administrative burden, incorporation limits the legal liability of your group's officers and members. It also provides a degree of respectability to your group in dealing with institutions—such as banks, the government, and foundations—and with the media and the public. And it is a necessary preliminary to seeking federal tax-exempt status, typically as a 501(c)(3) organization. (The primary advantages of attaining 501(c)(3) status are that it provides a further degree of respectability to your group and that donations to your group are then tax-deductible. A possible disadvantage is that such organizations—though not their individual members—are prohibited from participating in political campaigning.) If you decide to pursue incorporation, seek a sympathetic lawyer to provide pro bono assistance, which will involve filing articles of incorporation with the state and drafting bylaws. In certain cases, it may be possible for your group to operate under the nonprofit status of a friendly preexisting nonprofit organization.

A vital role for permanent state-level Citizens for Science groups is monitoring state legislatures for bills that seek to compromise the teaching of evolution. Assign members to monitor both the legislature's Web site and the state media, perhaps on a rotating basis. If any antievolution legislation is introduced, develop a set of talking points that explains what in particular is objectionable about the bill. If the bill is referred to committee (usually, but not always, the education committee), have members of your group contact members of the committee, particularly those who represent them, to explain the problems with the bill. If the committee conducts hearings on the bill, try to arrange for members of your group to testify against it. Throughout the process, continue to monitor the legislature's Web site for the progress of the bill and the media for commentary. If the committee passes the bill, have members of your group contact their legislators to explain the problems with the bill before it is returned for a floor vote. In general, remember that state legislators listen to their constituents, and the human touch is often useful: a friendly personal visit is preferable to a phone call; a polite personal letter is preferable to a form e-mail.

Permanent state-level Citizens for Science groups also have a vital role to play in the statewide textbook adoption process and the development of state science standards. As part of understanding the system, investigate the policies governing these processes—the state department of education is likely to have relevant information—and identify the opportunities for official participation and public comment. Your group is likely to include scientists and science educators who are eminently qualified to assess textbooks and to develop science standards; they ought to apply to be on the relevant committees, particularly because creationists—who recognize the impact of textbooks and state science standards on evolution education—are doing so. Ensure that your group avails itself of any opportunity for public comment during the process. If there are attempts to derail the process in order to impugn evolution—whether in the legislature (as in Minnesota in 2004, during the revision of its state science standards) or by the state board of education (as in Texas in 2003, during the biology textbook adoption process)—ensure that your group protests.

PITFALLS AND PROSPECTS

There are likely to be temptations to revise the goal of your group. For example, it is not uncommon for a controversy over evolution education in a community to be part and parcel of a broader controversy involving attempts of the religious right to expand its influence. If so, you might feel the need for your group to work against the religious right in general. Undertaking such a revision may be appropriate; the risk is that doing so might alienate those of your actual or potential members who are not concerned about the other issues. Similarly, it is not uncommon for atheists, agnostics, and humanists—who, naturally, are especially concerned about religious intrusions on the public schools—to assume a leadership role in your group. If so, they might argue that your group ought to be working against religion in general—thus reinforcing the second pillar of creationism! From the point of view of defending evolution education, such a revision of your group's goal is not going to be helpful. If members of your group are arguing that science unequivocally supports atheism and that your group should behave accordingly, remind them (in the words of the physical anthropologist Matt Cartmill), "It's an honorable belief, but it isn't a research finding."[14]

Keep in mind that your creationist opponents are in all likelihood organizing in the same way that you are: recruiting members from among their likely supporters, seeking resources and advice from state and national organizations, facilitating communications among their members, compiling talking points, monitoring the media and the policymaking bodies for developments, and so on. Have a member of your group observe their activities as far as possible. The goal is not to infiltrate their organization under false pretenses but simply to pay close attention to their public activities. Attend their meetings and collect their literature; subscribe to their public e-mail lists and visit their Web sites; keep track of their public utterances, in person and in print. Save everything, since at the beginning of a controversy creationists are often incautious, and any revealing comment on their part is potential ammunition for you down the road. (In Dover, Pennsylvania, for example, a school board member reportedly argued for the antievolution policy by declaiming, "This country wasn't founded on Muslim beliefs or evolution. This country was founded on Christianity, and our students should be taught as such"—a remark that came back to haunt him during the trial.) Assume that your creationist opponents are repaying the favor and that your activities, too, are constantly under scrutiny.

Perhaps the greatest challenge to any group is the risk of burnout. Defending the teaching of evolution is a difficult, lengthy, and often thankless task, especially when, as in small towns, the community is often divided by rancor and acrimony over the controversy. In Darby, Montana, for example, where a controversy over evolution education was prominent in 2004, activist Victoria Clark reports hearing comments like "The florist didn't deliver when she saw my name on the bill" and "My daughter stormed out of the classroom to avoid more trouble." She adds, "I even had a friend stop by the house and tell me that a fellow parishioner had asked her why I was leading up the religious education program at our church 'if I didn't believe in God.' "[15] If you have followed the advice for organizing, building coalitions, and dividing the labor, the likelihood of burnout will be diminished. Still, try to have what fun you can—events that foster camaraderie, even just a meal or a cup of coffee together, are always welcome. When the threat to evolution education is no longer imminent, feel free to relax, but make sure that the infrastructure for effective activism remains in place. Creationists are resilient, so it is unwise to assume that any victory for you is final.

Spreading the Word

Effective communication is critical to any successful defense of evolution. Whatever the format, a few basic principles are essential.

- DO YOUR HOMEWORK. *Understand the issues in the creationism/evolution controversy in general and in the local controversy in particular. You don't have to be experts—your opponents will not be—but a basic knowledge is essential.*

- BE CONCISE, CLEAR, AND TO THE POINT. *It is easy to become enmeshed in the complexities of the evolution/creationism controversy (particularly if you are a scientist). But your audience or readership probably is not as familiar with the controversy as you are. The acronymic advice KISS—Keep It Simple, Stupid—is deservedly famous.*

- BE CIVIL, CONFIDENT, AND CALM. *If your audience or readership perceives you as unpleasant, uncertain, or angry, the effect of your presentation is going to be lost, no matter how brilliant its content.*

- GET ON THE WEB. *For communication and outreach, the Internet is now essential. Developing a Web site for your group, even a basic, no-frills, purely text Web site, ought to be a top priority for your communications committee. Your group's Web site ought to complement and expand your group's communications in print, in public, and with the press.*

IN PRINT

Letters to the editor are a mainstay of any grassroots campaign. The point of writing is not to convince creationists that they are in error but to influence public opinion and inform policymakers. Before writing, ascertain what the newspaper's word limit for letters to the editor is—a limit of around two hundred words is not uncommon—and compose your letter accordingly. Effective letters are succinct, usually focusing on only one or a few central points, and often humorous. If you are responding to a previous creationist letter or essay, avoid the temptation to answer every single claim: it is often sufficient to expose the most glaring and absurd error, to invoke a relevant authority (such as the National Academy of Sciences on the evidence for evolution or the National Science Teachers Association on the importance of evolution education), or to address the issue in general terms. Play to your strengths: if you have special expertise—as a scientist, teacher, or member of the clergy, for example—invoke it in your letter. The editor is not likely to want to publish multiple letters by the same author or on the same topic, so coordinate with your fellow activists. Be persistent.

When what you have to say is simply too complicated to convey in a letter to the editor, it is time for the op-ed (named for its position opposite the editorial page). Op-ed essays are not only generally longer—in the range of around seven hundred words—but also generally expected to be better; the *New York Times* explains that its op-ed editors look for "timeliness, ingenuity, strength of argument, freshness of opinion, clear writing and newsworthiness," adding, "Personal experiences and first-person narrative can be great, particularly when they're in service to a larger idea. So is humor, when it's funny."[16] Especially because a good op-ed takes time and effort to compose, you will generally want to describe the prospective op-ed to the editor of your local newspaper and ask whether such a piece would be welcome before you start writing. Because the op-ed section is intended to complement, rather than repeat, the editorial position of the newspaper, try to present a different perspective on the issue at hand. As with letters to the editor, if you have special expertise, invoke it in your op-ed. And again, be persistent.

At any public event, distribute flyers or leaflets. Producing these is a task for a subcommittee of your group, ideally including someone who writes clearly and concisely and someone who is familiar with graphic design or typesetting. The goal is to produce flyers that are attractive, accurate, and easy to understand. There is no need for the text of the flyers to be original: resources from various national organizations can be adapted to fit the circumstances of your local situation. Remember to keep it simple: there is no sense in trying to cram a complete treatment of the creationism/evolution controversy on an 8½ x 11 inch sheet of paper, especially when it is always possible to include the titles of a few relevant books or a few useful Web sites instead. Be sure to include the address of your group's own Web site, or alternative contact information, on the flyers—prospective members need to be encouraged to join at every opportunity. Arrange to have the flyers posted on your organization's Web site as well, and encourage sympathizers to download, reprint, and distribute them.

IN PUBLIC

Attendance at public meetings of a school board or a legislative committee is vital; elected officials tend to gauge the level of public support for a policy by the number of people who actually attend the meeting. Enlist as many people as possible to attend such meetings; scatter throughout the audience (so as not to resemble a bloc) and applaud those on your side, if permitted. Consider wearing T-shirts or buttons with appropriate slogans to convey your unity. There is usually a limited amount of time available to comment or testify, so divide topics among the members of your group to ensure that every important point is covered. When speaking or testifying, emphasize why you care: parents want their children to receive a solid science education, teachers have a professional duty to teach in accordance with the scientific consensus, employers need to have scientifically literate employees, and so on. It is effective, if it can be arranged, to reserve a speaker from your group to get the last word in—such a cleanup hitter is able either to counter a creationist argument that you failed to anticipate or to wrap up your case. Submit your testimony in writing as well—someone may read it!

Supporters of evolution education are often challenged to engage in formal public debates on the scientific legitimacy of evolution and creationism, usually with creationists who specialize in such debates (such as the ubiquitous Kent Hovind of Creation Science Evangelism). Experience suggests that, as far as defending evolution is concerned, such debates are counterproductive: they presuppose and validate the false idea that creationism is a scientifically credible rival to evolution. Debates confuse the public about evolution and the nature of science; they increase the membership and swell the coffers of their creationist sponsors; and they fuel local enthusiasm for creationism, thereby contributing to pressure on local teachers to downplay evolution or even teach creationism. Faced with a challenge to engage in such a debate, politely decline. Consider a counteroffer of a written debate to be posted on the Web, which allows unlimited time and space, including the opportunity for documentation and references that would be impractical in oral debates. It is rare for a creationist to be willing to debate in such a format, which rewards science rather than showmanship.

Rather than participate in a rigged creationist debate, hold your own public education event (in a school, library, or college venue, perhaps in collaboration with other groups). Such forums help to attract attention to your group—be sure to alert the local media—as well as to the issue. Invite only panelists who support evolution education—the point of the event is to present, not debate, the facts, and creationists have plenty of events of their own in any case. A diverse panel is desirable: consider inviting a scientist, a member of the clergy, a high school biology teacher, a local education reporter, a member of the school board, a lawyer specializing in constitutional law, a philosopher of science, and so forth. Ask them to present brief statements expressing their perspectives on the issue. If you are able to recruit a national figure to speak at the event, so much the better! Have a member of your group moderate, introducing the panelists and managing a brief question-and-answer period. Use the opportunity to disseminate your group's flyers and leaflets, to build your group's mailing list, and (if permitted by the venue of the event) to take donations as well.

WITH THE PRESS

In any local creationism/evolution controversy, most of the public is going to learn about it from the local media. The mere fact that your group exists is useful for dealing with the press, since journalists like to be able to quote a representative of a formally organized group rather than just a random citizen. Members of your group can cultivate journalists, particularly education or science reporters, by, for example, dropping a friendly note to reporters who write on the controversy, commenting on their stories and offering their help for the future.[17] Professors in your group can ask their university or college press offices to list them as experts on relevant topics. Additionally, your group's communication committee may wish to designate a spokesperson or media relations person—who might or might not be the leader of the group—to deal with the press. Such a spokesperson ought to be knowledgeable and prepared, of course, although it's not inappropriate to say, "I don't know; let me check and get back to you." A spokesperson also ought to be friendly, persuasive, and, above all, accessible: having a spokesperson who isn't available to talk to a journalist facing a deadline is like not having a spokesperson at all.

The mere existence of your group also confers a degree of respectability on your press releases, which provide you with a way to try to steer press coverage. It is important not to overuse press releases, on the principle of the boy who cried wolf, so issue only press releases that are timely and topical. If they are not announcing a newsworthy event, such as a forum sponsored by your group, they should at least relate to a current newsworthy event, such as a public figure's statement on evolution education on which your group is commenting. As always, provide the information simply and clearly. In addition, help the reporters by providing a distinctive angle on the story, a snappy headline, and a few memorable quotes from prominent members of your group. Include a reason why the average reader ought to care—emphasizing the possible economic consequences of antievolution policies is often a winner. Although there are services that will compose and disseminate press releases for a fee, it is probably best for a grassroots organization to handle the process itself: hand-deliver, fax, and e-mail your press release to the local media outlets, particularly to friendly reporters.

Radio and television present challenges of their own, and it is important for your spokesperson to be prepared to be suitably articulate and telegenic if it is likely that such media opportunities are in the offing. The best format for you is the interview, in which your spokesperson will have the chance to engage in a substantive conversation with the show's host. The most frequent format is the news story, in which your spokesperson's contribution is limited to a few sentences, which themselves may be subject to editing. Here it is best to make the central points clearly, simply, and forcefully. The worst format—often on talk radio—is the debate, whether against a creationist guest (or even host) on the show or initiated by the show's callers. Such a debate format is potentially as counterproductive as a formal public debate, so be wary of accepting the offer to participate. In any format, if the show takes calls, encourage members of your group to listen and to offer supportive commentary.

In the Meantime

Even in the absence of a controversy, there are ways in which you can support evolution education in your local community. Urge educational policymakers—administrators in the local school district and the state department of education, members of the local and state school boards, legislators—to retain and expand the coverage of evolution and related concepts in state standards, textbooks, and local curricula. Donate books, videos, and instructional materials about evolution to your local public schools—for example, what biology teacher wouldn't like a set of replica fossil hominid skulls? Similarly, donate books and videos about evolution to both school and public libraries. (It is wise to ascertain beforehand that your donations are going to be welcome, however, since otherwise they will appear ignominiously in the Friends of the Library book sale.) If you attend a church or synagogue, discuss the possibility of adult religious education projects focusing on evolution with your priest, minister, or rabbi. Encourage and support evolution education in informal learning environments, such as natural history museums and science centers, parks and zoos and aquaria, and at Darwin Day celebrations.[18] Also encourage and support science education in the media—ask your local PBS station to rerun the 2001 series *Evolution*, for example, or the 2004 series *Origins*.[19]

If you are a parent, you have a special concern. Indeed, it is your child's understanding of science that is at stake. It behooves you to discuss activities and homework from science class with your child, to ensure that evolution is being taught and being taught appropriately. If so, let your child's science teachers know that they have your full support when it comes to teaching about evolution. Consider encouraging and assisting them to do so—by, for example, arranging field trips to informal learning environments, such as science centers and natural history museums, with scientifically accurate and age-appropriate exhibits about evolution. Be a full participant in your child's education: volunteer at the school; attend parent-teacher nights, open houses, and meetings of the PTA; be active in school board meetings, programs, and campaigns. (It isn't only evolution education that benefits from your doing so.)

If you are a professional science educator, scientist, or member of the clergy, you have further opportunities to support evolution education. As a science teacher, for example, you are on the front lines of the battle over evolution education. Arm yourself accordingly by keeping up on the latest developments both in the evolutionary sciences and in effective science pedagogy. If you are a new teacher, seek advice and support from your experienced colleagues; if you are a veteran teacher, provide advice and support to your novice colleagues. Read professional journals such as *The Science Teacher* and *The American Biology Teacher;* attend the conferences of the National Association of Biology Teachers and the National Science Teachers Association and support their affiliate organizations in your state; avail yourself of online resources such as those provided by the Evolution and the Nature of Science Institutes and the University of California Museum of Paleontology's Understanding Evolution project; work with your colleagues to develop or publicize workshops and in-service units about evolution and take advantage of them yourself at every opportunity.[20]

If you are a scientist, offer to speak on scientific topics relevant to education to school classes, civic groups, and church groups. If you teach at a college or university, incorporate evolutionary material as appropriate in your courses and encourage the creation of informal opportunities for evolution education, such as public lectures and museum exhibits. If you work in a museum, zoo, or other informal learning environment, incorporate evolution as appropriate in signage, docent education, and exhibit interpretation. Work through your professional societies as well, encouraging them to issue position statements supporting evolution education and to provide similar support as needed. (For instance, a friend-of-the-court brief supporting the decision in *Selman v. Cobb County* was submitted by fifty-six scientific societies, including the National Academy of Sciences and the American Association for the Advancement of Science.) Urge them to publish relevant articles and reviews in their journals and newsletters and on their Web sites. At conferences, organize sessions on evolution education for the attendees and provide workshops about evolution education for the local teachers.[21] Help professional scientific societies to help their members support evolution education in their local communities.

If you are a member of the clergy, you have a strong reason to want to defuse what is popularly regarded as a conflict between science and religion over evolution. Explore for yourself the rich and expanding theological literature on science and religion in general, perhaps with the aid of scientists and science educators in your congregation, and share the fruits of your research. Attempt to engage your congregations about the compatibility of evolution with your shared religious faith in sermons or in adult religious education projects.[22] Encourage your church or your denomination to produce educational resources about evolution and religion, such as the Episcopal Church's A Catechism of Creation,[23] and even to take a formal stand in support of evolution education. Bear in mind that defending the teaching of evolution is part of defending the separation of church and state, which—as its drafter James Madison predicted—is central to preserving the integrity and vitality of religion in the United States.

The suggestions offered here for defending and promoting the teaching of evolution in the public schools are by no means exhaustive. Opportunities abound for thoughtful, dedicated, and creative activists to use their unique set of talents, interests, and resources to make a difference. (Who would have thought, for example, that there was a biologist-cum-cartoonist? Yet there is, and Jay Hosler's graphic novel *The Sandwalk Adventures,* in which Darwin explains the rudiments of evolution to a follicle mite living in his left eyebrow, is not only graphically appealing and witty, scientifically accurate and pedagogically sophisticated, but it's also likely to have a greater influence on its adolescent readership than any number of earnest letters to the editor.[24]) By whatever means, defending the teaching of evolution in the public schools is essential for the continued scientific literacy of our nation's children. As Stephen Jay Gould once wrote, "Evolution is not a peripheral subject but the central organizing principle of all biological science. No one who has not read the Bible or the Bard can be considered educated in Western traditions; so no one ignorant of evolution can understand science."[25] If you care about the future of scientific literacy, it is time for you to help.

Notes

1. The Once and Future Intelligent Design

1. Edward J. Larson, *Trial and Error: The American Controversy over Creation and Evolution*, 3rd ed. (New York: Oxford University Press, 2003).

2. Karen Armstrong, *The Battle for God: A History of Fundamentalism* (New York: Ballantine Books, 2000).

3. Armstrong, *The Battle for God;* George M. Marsden, *Fundamentalism and American Culture: The Shaping of Twentieth-Century Evangelicalism, 1870–1925* (New York: Oxford University Press, 1980).

4. S. J. Holmes, "Proposed Laws Against the Teaching of Evolution," *Bulletin of the American Association of University Professors* 13, no. 8 (1927): 549–554.

5. Judith V. Grabiner and Peter D. Miller, "Effects of the Scopes Trial," *Science* 185, no. 4154 (1974): 832–837.

6. Larson, *Trial and Error.*

7. Arnold B. Grobman, "National Standards," *American Biology Teacher* 60, no. 10 (October 1998): 562.

8. Gerald Skoog, "Does Creationism Belong in the Biology Curriculum?," *American Biology Teacher* 40, no. 1 (January 1978): 23–29.

9. John C. Whitcomb and Henry M. Morris, *The Genesis Flood: The Biblical Record and Its Scientific Implications* (Phillipsburg, NJ: Presbyterian and Reformed, 1961).

10. Ronald L. Numbers, *The Creationists* (New York: Alfred A. Knopf, 1992).

11. Henry Morris, "Why ICR—and Why Now?," *Impact* 337 (July 2001): 4.

12. Edward J. Larson, *Summer for the Gods: The Scopes Trial and America's Continuing Debate over Science and Religion* (New York: Basic Books, 1997).

13. Randy Moore, "Thanking Susan Epperson," *American Biology Teacher* 60, no. 11 (November/December 1998): 642–646.

14. Wendell R. Bird, "Resolution for Balanced Presentation of Evolution and Scientific Creationism," *Impact* 71 (May 1979): 4.

15. Anonymous, "Creation Science and the Local School District," *Impact* 67 (January 1979): 4.

16. For Alabama through Washington, see Wayne Moyer, "Legislative Initiatives," *Scientific Integrity* (June 1981): 1–4; for Georgia through Wisconsin, see Anonymous, "Update on Creation Bills and Resolutions," *Creation/Evolution* 2, no. 1 (1998): 1–44; for the remaining states, consult various Memoranda to Liaisons for Committees of Correspondence, available in NCSE's archives.

17. Anonymous, Act 590 of 1981, reprinted in *Creationism, Science, and the Law*, ed. M. C. La Follette (Cambridge, MA: MIT Press, 1983).

18. Charles B. Thaxton, Walter L. Bradley, and Roger L. Olsen, *The Mystery of Life's Origin: Reassessing Current Theories* (New York: Philosophical Library, 1984).

19. Fred Hoyle and Chandra Wickramasinghe, *Diseases from Space* (New York: Harper and Row, 1979).

20. Percival W. Davis and Dean H. Kenyon, *Of Pandas and People*, 2nd ed. (Dallas: Foundation for Thought and Ethics, 1993).

21. Henry M. Morris, "The Design Revelation," *Back to Genesis* 194 (February 2005): 2–3.

22. Duane T. Gish, *Evolution: The Fossils Still Say No!* (San Diego: Institute for Creation Research, 1985).

23. Jonathan Wells, *Icons of Evolution: Science or Myth?* (Washington, DC: Regnery, 2000).

24. Wayne Frair and Percival Davis, *A Case for Creation*, 3rd ed. (Chicago: Moody Press, 1983); Stephen C. Meyer, "The Origin of Biological Information and the Higher Taxonomic Categories," *Proceedings of the Biological Society of Washington* 117, no. 2 (2004): 213–239.

25. William Dembski, *The Design Inference: Eliminating Chance Through Small Probabilities* (New York: Cambridge University Press, 1998).

26. Wayne Frair, "Baraminology—Classification of Related Organisms," *Creation Research Quarterly* 37, no. 2 (2000): 82–91.

27. William Dembski, "Why President Bush Got It Right about Intelligent Design," blog post dated August 4, 2005; available online at www.uncommondescent.com/index.php/archives/222; Henry Morris, *That You Might Believe* (Chicago: Good News, 1946).

28. Phillip E. Johnson, *Darwin on Trial* (Washington, DC: Regnery Gateway, 1991).

29. Ken Walker, "Young-Earth Theory Gains Advocates," *Christianity Today* 42, no. 5 (1998): 24.

30. Bruce Chapman, press release (Seattle: Discovery Institute, Oct. 10, 1996).

31. William Dembski, *Intelligent Design: The Bridge Between Science and Theology* (Downers Grove, IL: InterVarsity Press, 1999), 13.

32. Barbara Forrest and Paul R. Gross, *Creationism's Trojan Horse* (New York: Oxford University Press, 2004).

33. David K. DeWolf, Stephen C. Meyer, and Mark E. DeForrest, *Intelligent Design in Public School Science Curricula: A Legal Guidebook* (Richardson, TX: The Foundation for Thought and Ethics, 1999).

2. Analyzing Critical Analysis: The Fallback Antievolutionist Strategy

1. For the history of creation science, see Ronald L. Numbers, *The Creationists: The Evolution of Scientific Creationism* (New York: Alfred A. Knopf, 1992). For the legal history of the issue, see Edward J. Larson, *Trial and Error: The American Controversy over Creation and Evolution*, 3rd ed. (New York: Oxford University Press, 2003). For the history of ID, see *Kitzmiller v. Dover Area School District, et al.* (December 20, 2005), available at www.pamd.uscourts.gov/kitzmiller/kitzmiller_342.pdf.

2. Wendell R. Bird and Institute for Creation Research staff, "The Supreme Court Decision and its Meaning," *Impact* 170 (August 1987), emphasis in the original.

3. The Discovery Institute CRSC has since relabeled itself the Center for Science and Culture, presumably to appear more secular. See NCSE (2002), "Evolving Banners at the Discovery Institute," www.ncseweb.org/resources/articles/8325_evolving_banners_at_the_discov_8_29_2002.asp. For the relationship of "The Wedge" to earlier DI documents, see Barbara Forrest and Paul R. Gross, *Creationism's Trojan Horse: The Wedge of Intelligent*

Design (New York: Oxford University Press, 2004), 15–33. A scan of the original Wedge document was put online in 2006 by the *Seattle Weekly* at www.seattleweekly.com/news/0605/discovery-wedge.php.

4. Jonathan Wells, *Icons of Evolution: Science or Myth?* (Washington, DC: Regnery Publishing, 2000). For an overview critique, see Kevin Padian and Alan D. Gishlick, "The Talented Mr. Wells," *The Quarterly Review of Biology* 77, no. 1 (2002): 33–37. For more details, see Alan D. Gishlick, "Icons of Evolution? Why Much of What Jonathan Wells Writes about Evolution Is Wrong," available at www.ncseweb.org/icons/, and various articles available at www.talkorigins.org/faqs/wells/.

5. See "NCSE Compilation on Santorum Amendment," www.ncseweb.org/resources/articles/7202_ncse_compilation_on_santorum_a_6_13_2002.asp. For summaries of recent controversies, see www.ncseweb.org/pressroom.asp.

6. Ohio Board of Education (2004), "L10H23 Lesson Plan: Critical Analysis of Evolution." See http://science2.marion.ohio-state.edu/ohioscience/lesson-plans.html.

7. Patricia Princehouse, Ohio Citizens for Science, personal communication.

8. Paul R. Gross et al., *The State of State Science Standards 2005* (Washington, DC: Thomas B. Fordham Institute, 2005).

9. Kansas Board of Education, "Kansas Science Education Standards, Approved November 8, 2005," available at www.ksde.org/outcomes/sciencestd.pdf. In this chapter, we quote from the Rationale, p. ii, and Standard 3: Life Science, pp. 71–83.

10. Transcripts of the hearings are available online at: www.talkorigins.org/faqs/kansas/kangaroo.html. Scientists properly qualified in the subject matter boycotted these hearings, arguing that they amounted to a kangaroo court since they were overseen by the three staunchest creationists on the board, aided by lawyer John Calvert of the Intelligent Design Network, and this group would devise the rules and serve as judge and jury.

11. The topic of statistical similarity of differing phylogenetic trees is discussed in detail, with a thorough review of the relevant scientific literature, by Douglas Theobald, "Statistical Support for Phylogenies." See www.talkorigins.org/faqs/comdesc/phylo.html#reliability.

12. The key paper is Jean-Renaud Boisserie, Fabrice Lihoreau, and Michel

Brunet, "The Position of Hippopotamidae Within Cetartiodactyla," *Proceedings of the National Academy of Sciences* 102, no. 5 (2005): 1537–1541; Robert Sanders provides a nontechnical account: "UC Berkeley, French Scientists Find Missing Link Between the Whale and Its Closest Relative, the Hippo." See www.berkeley.edu/news/media/releases/2005/01/24_hippos.shtml.

13. Stephen Jay Gould, "Evolution as Fact and Theory," in *Hen's Teeth and Horse's Toes: Further Reflections in Natural History* (New York: W. W. Norton, 1983), 258–260.

14. For shell borers, see p. 301 of Stefan Bengtson, "Origins and Early Evolution of Predation," *The Fossil Record of Predation,* eds. Michal Kowalewski and Patricia H. Kelley, *The Paleontological Society Papers* 8 (2002): 289–317. The Cambrian Explosion was not an instantaneous event: "The oldest trace fossils are approximately 550 Ma, giving a period of at least 30 Ma before the appearance of trilobites." From p. 161, Graham E. Budd, "The Cambrian Fossil Record and the Origin of the Phyla," *Integrative and Comparative Biology* 43 (2003): 157–165.

15. Graham E. Budd, "The Cambrian Fossil Record and the Origin of the Phyla," p. 159, emphasis ours.

16. Degan G. Shu, Simon Conway Morris, Jian Han, Zhifei F. Zhang, and Jianni N. Liu, "Ancestral Echinoderms from the Chengjiang Deposits of China," *Nature* 430, no. 6998 (2004): 422–428.

17. David Fitch and Walter Sudhaus, "One Small Step for Worms, One Giant Leap for 'Bauplan'?," *Evolution and Development* 4, no. 4 (2002): 243–246. The Conway Morris quote is on p. 170 of Simon Conway Morris, *The Crucible of Creation: The Burgess Shale and the Rise of Animals* (New York: Oxford University Press, 1998). For a discussion of this and related research, see the post "Down with Phyla!" at the Panda's Thumb blog: www.pandasthumb.org/archives/2005/04/down_with_phyla_1.html.

18. Alan D. Gishlick, "Icons of Evolution?" See especially figure 8, comparing embryo photos, and figure 10, comparing textbooks, at: www.ncseweb.org/icons/figures.html.

19. Michael K. Richardson, James Hanken, Lynne Selwood, Glenda M. Wright, Robert J. Richards, Claude Pieau, and Albert Raynaud, "Letter," *Science* 280, no. 5366 (1998): 983.

20. For Kenyon's affidavit, see www.talkorigins.org/faqs/edwards-v-aguillard/kenyon.html. The Meyer quote is from p. 68 of Stephen C. Meyer, "The Methodological Equivalence of Design and Descent: Can There Be a

Scientific 'Theory of Creation'?," in J. P. Moreland, ed., *The Creation Hypothesis: Scientific Evidence for an Intelligent Designer* (Downers Grove, IL: InterVarsity Press, 1994), 67–112.

21. Manyuan Long, Esther Betrán, Kevin Thornton, and Wen Wang, "The Origin of New Genes: Glimpses from the Young and Old," *Nature Reviews Genetics* 4, no. 1 (2003): 865–875. For Manyuan Long's homepage, see http://pondside.uchicago.edu/ceb/faculty/Long.html.

22. Michael J. Behe, *Darwin's Black Box* (New York: The Free Press, 1996), 138.

23. For a discussion of the role of immunology in the *Kitzmiller v. Dover* trial, containing references to the literature surveyed here, see Andrea Bottaro, Matt A. Inlay, and Nicholas J. Matzke, "Immunology in the Spotlight at the Dover 'Intelligent Design' Trial," *Nature Immunology* 7, no. 5 (2006): 433–435.

24. Frank Lewis Marsh, "The Form and Structure of Living Things," *Creation Research Society Quarterly* 6, no. 1 (1969): 13–25. The quoted passages are from pp. 18–20, emphases in the original.

25. Ian McDowell, "Questions for Evolutionists," *Creation Research Society Quarterly* 7, no. 3 (1970): 182–183; Bolton Davidheiser, "The Human Quest: A New Look at Science and Christian Faith by Richard H. Bube" (review), *Creation Research Society Quarterly* 9, no. 2 (1972): 141–142. In addition to the "it's just microevolution" pseudoargument about peppered moths, teachers regularly confront exaggerated new challenges to the peppered moth example. For a recent informed review of the situation, see Jim Mallet, "The Peppered Moth: A Black and White Story After All," *Genetics Society News* 50 (2004): 34–38.

26. Douglas J. Futuyma, *Evolutionary Biology* (Sunderland, MA: Sinauer Associates, 1998); Stephen Jay Gould, "Macroevolution," in Mark Pagel, ed., *Encyclopedia of Evolution* (Oxford: Oxford University Press, 2002), vol. 1: E-23–E-28; Richard Dawkins, *The Ancestor's Tale: A Pilgrimage to the Dawn of Evolution* (Boston: Houghton Mifflin, 2004).

27. Kurt P. Wise, "The Evolution of Creationist Perspective on the Fossil Equid Series," *Geological Society of America 2003 Annual Meeting Abstracts with Programs* 35, no. 6 (2003): 610; David P. Cavanaugh and Todd Charles Wood, "A Baraminological Analysis of the Tribe Heliantheae sensu lato (Asteraceae) Using Analysis of Pattern (ANOPA)," *Occasional Papers of the Baraminology Study Group* 1 (2002): 1–11; Wayne P. Armstrong, "Sunflower Family (Asteraceae): The Largest Plant Family on Earth," *Wayne's Word: An Online Textbook of Natural History* 9, no. 3 (2000); Ronald L. Numbers, "Ironic Heresy: How Young-Earth Creationists Came to Embrace Rapid Microevo-

lution by Means of Natural Selection," in Abigail Lustig, Robert J. Richards, and Michael Ruse, eds., *Darwinian Heresies* (Cambridge: Cambridge University Press, 2004), 84–100.

28. Phillip E. Johnson, *Darwin on Trial* (Downers Grove, IL: InterVarsity Press, 1991), 68.

29. The *Pandas* quotes are from Percival William Davis and Dean Kenyon, *Of Pandas and People: The Central Question of Biological Origins*, 2nd ed. (Dallas: Foundation for Thought and Ethics, 1993), 11, 61, emphasis in original. The *Bible-Science Newsletter* articles by Nancy R. Pearcey are "Bible Study: Which Is More Scientific: 'Kinds' or 'Species'?," *Bible-Science Newsletter* 27, no. 5 (1989): 9–10; "Echo of Evolution? The Revolution in Molecular Biology," *Bible-Science Newsletter* 27, no. 12 (1989): 7–11; "Of Fins and Fingers: Patterns in Living Things," *Bible-Science Newsletter* 27, no. 5 (1989): 6–9; "What Species of Species?—or, Darwin and the Origin of *What?*," *Bible-Science Newsletter* 27, no. 6 (1989): 7–9.

30. Discovery Institute, "The Scientific Controversy over Whether Microevolution Can Account for Macroevolution," available at www.discovery.org/scripts/viewDB/filesDB-download.php?id=118.

31. Steve Abrams, "Science Standards Aren't about Religion," *Wichita Eagle*, November 15, 2005.

3. Theology, Religion, and Intelligent Design

1. William Paley, *Natural Theology: or, Evidences for the Existence and Attributes of the Deity, Collected from the Appearances of Nature* (London: R. Faulder, 1802).

2. Richard Dawkins, *The Blind Watchmaker: Why the Evidence of Evolution Reveals a Universe without Design* (New York: W. W. Norton, 1987).

3. Phillip E. Johnson, *Darwin on Trial* (Washington, DC: Regnery Gateway, 1991), 69.

4. Henry M. Morris, *Scientific Creationism* (Green Forest, AR: Master Books, 1974, 1985); Duane T. Gish, *Evolution: The Fossils Still Say No!* (El Cajon, CA: Institute for Creation Research, 1985).

5. Michael Behe, *Darwin's Black Box: The Biochemical Challenge to Evolution* (New York: Touchstone/Simon Schuster, 1996), 39.

6. William Dembski, *No Free Lunch: Why Specified Complexity Cannot Be Purchased without Intelligence* (Lanham, MD: Rowman & Littlefield, 2002), 12.

7. Ibid., 15.

8. Johnson, *Darwin on Trial,* 14.

9. Ibid., 42.

10. For a description of scientism, see Ted Peters, "Science and Theology: Toward Consonance," in Ted Peters, ed., *Science and Theology: The New Consonance* (Boulder, CO: Westview Press, 1998), 11–39.

11. Terry Gray, "The Mistrial of Evolution: A Review of Phillip E. Johnson's *Darwin On Trial,*" online version, at www.asa3.org/gray/evolution_trial/dotreview.html.

12. Johnson, *Darwin on Trial,* 20.

13. Michael Ruse, *Taking Darwin Seriously* (Amherst, NY: Prometheus Books, 1998), 24.

14. Kenneth Miller, *Finding Darwin's God* (New York: Cliff Street Books, 1999), 160.

15. Dembski, *No Free Lunch,* xiii–xiv.

16. Robert John Russell, "Divine Action and Quantum Mechanics: A Fresh Assessment," and William Stoeger, "Epistemological and Ontological Issues Arising from Quantum Mechanics," both in Robert John Russell, Philip Clayton, Kirk Wegter-McNelly, and John Polkinghorne, eds., *Quantum Mechanics: Scientific Perspectives on Divine Action* (Vatican Observatory Press, distributed by University of Notre Dame Press, 2001): 293–328 (Russell), 81–98 (Stoeger).

17. Perhaps the most articulate defender of the value of evolutionary theory for generating progressive research has been Stephen Jay Gould. See his *The Structure of Evolutionary Theory* (Cambridge, MA: Harvard University Press, 2002) and the analysis by Francisco Ayala, "The Structure of Evolutionary Theory: On Stephen Jay Gould's Monumental Masterpiece," *Theology and Science* 3, no. 1 (2005): 97–117.

18. Huston Smith, *The World's Religions* (San Francisco: Harper, rev. ed., 1991).

19. Huston Smith, *Why Religion Matters* (San Francisco: Harper, 2001), 176–178.

20. See www.religionandecology.org.

21. See Web site for "Judaism and Evolution," http://en.wikipedia.org/wiki/Jewish_creationism.

22. Muzaffar Iqbal, *Islam and Science* (Aldershot, UK: Ashgate, 2002), xvii.

23. Medhi Golshani, "Creation in the Islamic Outlook and in Modern Cosmology," in *God, Life, and the Cosmos: Christian and Islamic Perspectives*, ed. Ted Peters, Muzaffar Iqbal, and Syed Nomanul Haq (Aldershot, UK: Ashgate, 2002), 224.

24. Iqbal, *Islam and Science*, 276.

25. T. O. Shanavas, *Creation and/or Evolution: An Islamic Perspective* (Philadelphia: XLIBRIS, 2005), 123.

26. Ibid., 124.

27. Ibid., 11.

28. Seyyed Hossein Nasr, *Religion and the Order of Nature* (Oxford and New York: Oxford University Press, 1996), 6.

29. Ibid., 146.

30. Harun Yahya, "The Theory of Evolution: A Unique Deception in the History of the World," available at www.harunyahya.com/articles/unique_deception_evolution.php. See also his book *The Evolution Deceit* (Istanbul: Okur Publishing, 2000).

4. From the Classroom to the Courtroom: Intelligent Design and the Constitution

1. *Kitzmiller v. Dover Area School District* (December 20, 2005), available at www.pamd.uscourts.gov/kitzmiller/kitzmiller_342.pdf.

2. Much of the discussion in this chapter is adopted or taken directly from several other articles I have written about intelligent design. Jay D. Wexler, "Of Pandas, People, and the First Amendment: The Constitutionality of Teaching Intelligent Design in the Public Schools," *Stanford Law Review* 49 (1997): 439–470; Jay D. Wexler, "Darwin, Design, and Disestablishment: Teaching the Evolution Controversy in Public Schools," *Vanderbilt Law Review* 56 (2003): 751–855; Jay D. Wexler, "Intelligent Design and the First Amendment: A Response," *Washington University Law Quarterly* 53 (forthcoming, 2006). In this chapter I consider primarily the constitutional ramifications of a school board that adopts a formal intelligent design policy, but it is also worth noting that many of the same arguments can be applied to an individual teacher who determines on his or her own to teach intelligent design as a legitimate scientific theory in the classroom.

3. *Lemon v. Kurtzman*, 403 U.S. 602 (1971).

4. *Allegheny, County of, v. ACLU*, 492 U.S. 573 (1989).

5. *Capitol Square Review and Advisory Board v. Pinette*, 515 U.S. 753, 780 (O'Connor, J., concurring).

6. *Epperson v. Arkansas*, 393 U.S. 97 (1968).

7. *Edwards v. Aguillard*, 482 U.S. 578 (1987).

8. The eight cases are *Edwards, Epperson, School District of Abington Township v. Schempp*, 374 U.S. 203 (1963); *Engel v. Vitale*, 370 U.S. 421 (1962); *Stone v. Graham*, 449 U.S. 39 (1980); *Wallace v. Jaffree*, 472 U.S. 38 (1985); *Lee v. Weisman*, 505 U.S. 577 (1992); and *Santa Fe Independent School District v. Doe*, 530 U.S. 290 (2000).

9. *Torcaso v. Watkins*, 367 U.S. 488, 495 n.11 (1961).

10. For example, this test was adopted by the Third Circuit Court of Appeals in *Africa v. Pennsylvania*, 662 F.2d 1025 (3rd Cir. 1981).

11. For example, see Francis J. Beckwith, *Law, Darwinism, and Public Education: The Establishment Clause and the Challenge of Intelligent Design* (Lanham, MD: Rowman & Littlefield, 2003), 152.

12. For an extended argument that schools should teach students about religion to prepare them for civic life in a nation (and world) populated by believers of many religious traditions, see Jay D. Wexler, "Preparing for the Clothed Public Square: Teaching About Religion, Civic Education, and the Constitution," *William and Mary Law Review* 43 (2002): 1159–1263.

13. An example is the program developed at the University of Wisconsin–Stout called "The Stout Science Program for Educators," which, according to its mission statement, "is designed to provide teachers with the tools necessary to improve student competencies in science and critical thinking . . . by using good science and pseudoscience (or false science) as an instructional tool." The program's Web site can be found at http://physics.uwstout.edu/stoutsci/.

14. For a description of a wide variety of creation stories, see David Leeming and Margaret Leeming, *A Dictionary of Creation Myths* (New York: Oxford University Press, 1994).

15. For the judge's account of this religious language, and the Wedge document in particular, see *Kitzmiller*, 24–35. A comprehensive account of this history can be found in Barbara Forrest and Paul R. Gross, *Creationism's Trojan Horse: The Wedge of Intelligent Design* (New York: Oxford University Press, 2004).

16. More on this point can be found in Jay D. Wexler, "Framing the Public Square," *Georgetown Law Journal* 91 (2002): 183–218.

17. For example: Michael W. McConnell, "Religion at a Crossroads," *University of Chicago Law Review* 59 (1990): 115, 154; Steven D. Smith, "Separation and the 'Secular': Reconstructing the Disestablishment Decision," *Texas Law Review* 67 (1989): 955–1032.

18. *Kitzmiller*, n.7.

19. Beckwith, *Law, Darwinism, and Public Education*, 76.

20. Nicholas P. Miller, "Life, the Universe and Everything Constitutional: Origins in the Public Schools," *Journal of Church and State* 43 (2001): 500.

21. *Keyhishian v. Board of Regents*, 385 U.S. 589 (1967).

22. *Pickering v. Board of Education*, 391 U.S. 563 (1968).

23. *Urofsky v. Gilmore*, 216 F.3d 401, 412 (4th Cir. 2000).

24. Ibid., 407.

25. Beckwith, *Law, Darwinism, and Public Education*, 76.

5. Evolution in the Classroom

1. National Academy of Sciences, *Science and Creationism: A View from the National Academy of Sciences*, 2nd ed. (Washington, DC: National Academy Press, 1999).

2. See www.nsta.org/pressroom&news_story_ID=50377. Although this was an informal, Web-based, unscientific poll, its results are consistent with those of formal scientific polls.

3. American Association for the Advancement of Science, 2002 AAAS Board Resolution on Intelligent Design Theory. See www.aaas.org/news/releases/2002/1106id2.shtml.

4. National Academy of Sciences, *Evolution in Hawaii* (Washington, DC: National Academy Press, 2004).

5. National Association of Biology Teachers, 2004, NABT's Statement on Teaching Evolution. See www.nabt.org/sub/position_statements/evolution/asp.

6. National Science Teachers Association, 2005, NSTA Position Statement:

The Teaching of Evolution. See http://nsta.org/positionstatement&psid=10&print=y.

7. National Science Teachers Association, 2005, NSTA Background Paper on the Use and Adoption of Textbooks in Science Teaching. See http://nsta.org/textbooks.

8. Association for Science Teacher Education, 2004, ASTE Position Statement: Science Teacher Preparation & Professional Development. See http://aste.chem.pitt.edu/.

9. National Science Teachers Association, 2004, NSTA Position Statement: Science Teacher Preparation. See www.nsta.org/positionstatement&psid=42.

10. Joel Cracraft and Roger W. Bybee, eds., *Evolutionary Science and Society: Educating a New Generation* (Washington, DC: American Institute of Biological Sciences, 2005).

11. *Kitzmiller v. Dover Area School District* (December 20, 2005), available at www.pamd.uscourts.gov/kitzmiller/kitzmiller_342.pdf.

12. For discussions of methodological naturalism, see chapter 12 of Eugenie C. Scott, *Evolution vs. Creationism* (Berkeley: University of California Press, 2005).

13. For help with these challenges and creationist student questions, see Brian Alters and Sandra Alters, *Defending Evolution in the Classroom: A Guide to the Creation/Evolution Controversy* (Boston: Jones and Bartlett, 2001).

14. National Research Council, *How Students Learn: Science in the Classroom* (Washington, DC: The National Academies Press, 2005).

15. Brian Alters and Craig E. Nelson, "Perspective: Teaching Evolution in Higher Education," *Evolution* 56, no. 10 (2002): 1891–1901. See also Brian Alters, *Teaching Biological Evolution in Higher Education: Methodological, Religious, and Non-Religious Issues* (Boston: Jones and Bartlett, 2005).

16. Stephen Jay Gould, *The Lying Stones of Marrakech* (New York: Harmony Books, 2000).

17. Howard Gardner, *Intelligence Reframed: Multiple Intelligences for the 21st Century* (New York: Basic Books, 1999).

6. Defending the Teaching of Evolution: Strategies and Tactics for Activists

1. The cited polls were conducted by CBS News in November 2004 and by *Newsweek* in December 2004. Note, however, that it is unclear whether these polls provide a sensitive assessment of public opinion. In a poll that offered respondents a wider array of choices, only 13 percent favored teaching creationism as a scientific theory along with evolution, and only 16 percent favored teaching creationism instead of evolution. The latter poll was conducted by DYG on behalf of the People for the American Way Foundation in November 1999; see www.pfaw.org/pfaw/dfiles/file_36.pdf. For useful discussions of public opinion polling with respect to evolution, see George Bishop, "The Religious Worldview and American Beliefs about Human Origins," *Public Perspective* 9, no. 5 (1998): 39–44; George Bishop, "'Intelligent Design': Illusions of an Informed Public," *Public Perspective* 14, no. 3 (2003): 40–42; Matthew C. Nisbet and Erik C. Nisbet, "Evolution and Intelligent Design: Understanding Public Opinion," *Geotimes* 50, no. 8 (2005): 28–33.

2. In controversies over evolution education, the word *evolution* often stands for a range of scientific disciplines, from cosmology and astronomy through geology and origin-of-life research to paleontology and anthropology, that are rejected in part or in whole by creationists. Strictly speaking, however, theories about the origin of the universe, the age of the Earth, and the origin of life are not part of the theory of evolution, although they, too, deserve to be taught free from religious intrusion. In modern biology, *evolution* refers primarily to what Darwin called "descent with modification"—living things have descended, with modification, from common ancestors—but it also refers by extension to the theories of the patterns and processes by which evolution occurs, which are also often under creationist attack. Evolution (in all of the senses above) is accepted on the basis of scientific evidence. To accept evolution is not to embrace a religion or subscribe to a philosophy, as creationists often contend (sometimes using the term *evolutionism*); since *belief* connotes faith, it is useful to talk about accepting evolution rather than believing it.

3. The so-called Santorum language, which was written by a proponent of intelligent design, added by Senator Rick Santorum (R–PA) as a sense of the Senate amendment to what was to become the No Child Left Behind Act, and removed from the bill by the joint conference committee, is the most prominent case in point. For details, see Glenn Branch and Eugenie C. Scott, "The Antievolution Law That Wasn't," *American Biology Teacher* 65, no. 3 (2003): 165–166.

4. See Jeffrey Weld and Jill C. McNew, "Attitudes Toward Evolution," *The Science Teacher* 66, no. 9 (1999): 26–31, and NSTA's press release at www.nsta.org/pressroom&news_story_ID=50377.

5. Eugenie C. Scott and Glenn Branch, "Evolution: What's Wrong with 'Teaching the Controversy,'" *Trends in Ecology and Evolution* 18, no. 10 (2003): 499–502.

6. Paul R. Gross, *The State of State Science Standards 2005* (Washington, DC: Thomas B. Fordham Institute, 2005), available also at www.edexcellence.net/institute/publication/publication.cfm?id=352, discusses the quality of state standards overall, with special attention to evolution. Also relevant are three articles by Sean Cavanagh—"Treatment of Evolution Inconsistent," *Education Week,* November 9, 2005, 20–21; "Evolution Theory Well Represented in Leading High School Textbooks," *Education Week,* December 7, 2005, 10; and "Many States Include Evolution Questions on Assessments," *Education Week*, December 7, 2005, 10–11—as well as a November 2005 research brief conducted by Christopher B. Swanson for *Education Week*'s publisher, the Editorial Projects in Education Research Center, on "Evolution in State Science Standards." All are available together at www.edweek.org/media/evolution1207.pdf.

7. For refutations of particular creationist claims about science, in addition to chapter 2, Mark Isaak, *The Counter-Creationism Handbook* (Westport, CT: Greenwood Press, 2005) is helpful. On the Web, the Talk Origins Archive at www.talkorigins.org is invaluable. NCSE's Project Steve, at www.ncseweb.org/article.asp?category=18, is a tongue-in-cheek response to the creationist lists of scientists who reject evolution. Moreover, such creationist lists clearly represent only the *opinion* of their signers. There is no original work in the peer-reviewed scientific research literature to support their view; see, for example, Barbara Forrest and Paul R. Gross, *Creationism's Trojan Horse* (New York: Oxford University Press, 2004), 43–47. Expect to hear the charge that articles providing evidence for creationism are censored; a dated but still useful article showing that proponents of creation science were not even *submitting* such articles is Eugenie C. Scott and Henry P. Cole, "The Elusive Scientific Basis of Creation 'Science,'" *The Quarterly Review of Biology* 60, no. 1 (1985): 21–29. A resource that debunks creationist misuse of the scientific literature is the Talk Origins Archive's Quote Mine Project, edited by John Pieret, at www.talkorigins.org/faqs/quotes/mine/project.html.

8. National Academy of Sciences, *Science and Creationism,* 2nd edition (Washington, DC: National Academies Press, 1999), 1–2.

9. There are a handful of vocal antievolutionists who appear to be driven more by contrariety than by piety, but their influence is negligible except when they are allied with, or invoked as allies by, creationists.

10. For John Paul II's 1996 statement and commentary by Edmund D. Pellegrino, Michael Ruse, Richard Dawkins, and Eugenie C. Scott, see "The Pope's Message on Evolution and Four Commentaries," *The Quarterly Review of Biology* 72, no. 4 (1997): 375–406.

11. See Michael Zimmerman's Clergy Letter Project, available at www.uwosh.edu/colleges/cols/clergy_project.htm.

12. For examples of the fallback strategy, see Wendell R. Bird and Institute for Creation Research staff, "The Supreme Court Decision and Its Meaning," *Impact* 170 (August 1987); Stephen C. Meyer, "Teach the Controversy," *Cincinnati Enquirer*, March 30, 2003. Meyer is director of the Discovery Institute's Center for Science and Culture, the institutional home of intelligent design creationism.

13. In dealing with a problem at the level of the individual classroom, you—and any allies that you are able to muster—ought to follow the chain of command from the teacher to the chair of the science department (if there is one) to the principal, and then to the school district (which often employs a science education specialist—a person whose interest is worth cultivating) and the school board. At every meeting and in every written communication, ask for a specific solution or response by a given date, thereby sending the message that you are not going to forget about the problem and that you are prepared to pursue the matter to the next level of the chain of command. Take notes during or immediately after any meeting, and insist on documentation and explanations of the claims offered by the teacher, school, and district. Document your own claims, especially with appeal to relevant authority (scientific, educational, and legal). Although such problems often are resolved quickly and quietly, out of the public view, there is always a chance of, and sometimes a necessity for, the problem to be publicly exposed: it then becomes a wider controversy in which the whole school district is embroiled, and it will be necessary for you to organize a group.

14. Matt Cartmill, "Oppressed by Evolution," *Discover* 19, no. 3 (March 1998): 78–83.

15. Victoria Clark, "Shall We Let Our Children Think?" *Reports of the National Center for Science Education* 24, no. 2 (2004): 10–11.

16. David Shipley, "And Now a Word from Op-Ed," *New York Times*, February 1, 2004.

17. A useful article, both to help you understand the challenges faced by the press in covering the creationism/evolution issue and to recommend to journalists facing those challenges, is Chris Mooney and Matthew C. Nisbet, "Undoing Darwin," *Columbia Journalism Review* 44, no. 3 (September/October 2005): 31–39.

18. Usually held on or near February 12, Darwin Day celebrations provide a marvelous opportunity not only to honor the life and work of Darwin but also to engage in public outreach about science, evolution, and the importance of evolution education. To find a Darwin Day event in your area, or to register your own event, see www.darwinday.org.

19. Indeed, because slick creationist programs such as *Unlocking the Mystery of Life* have occasionally found their way into PBS distribution networks, it is worth educating the programming staff at your local PBS station about the need to ensure that the science programs they broadcast are indeed legitimate.

20. *The Science Teacher* and *The American Biology Teacher* are published by the National Science Teachers Association (www.nsta.org) and the National Association of Biology Teachers (www.nabt.org), respectively. The Evolution and the Nature of Science Institutes Web site is at www.indiana.edu/~ensiweb/home.html. The University of California Museum of Paleontology's Understanding Evolution Web site is at http://evolution.berkeley.edu.

21. A blueprint for organizing such workshops is available at www.ucmp.berkeley.edu/ncte/twb/.

22. The 197th anniversary of Darwin's birthday, February 12, 2006, was also the first Evolution Sunday. More than 450 churches across the country conducted sermons or adult religious education programs affirming the compatibility of science with their religious faith. For details, see www.evolutionsunday.org.

23. Available at www.dfms.org/19021_58393_ENG_HTM.htm.

24. Jay Hosler, *The Sandwalk Adventures* (Columbus, OH: Active Synapse, 2003).

25. Stephen Jay Gould, "Dorothy, It's Really Oz: A Pro-Creationist Decision in Kansas Is More Than a Blow Against Darwin," *Time*, August 23, 1999, 59.

Contributors

Brian Alters is the Tomlinson Chair in Science Education at McGill University; Director, Tomlinson University Science Education Project; and founder and director of the Evolution Education Research Centre. He also has held an appointment in the Harvard College Observatory at Harvard University for ten years. He is a global leader in the field of evolution education, coauthor of *Defending Evolution: A Guide to the Evolution/Creation Controversy*, and author of *Teaching Biological Evolution in Higher Education*. In 2005, he testified as an expert witness for the plaintiffs in *Kitzmiller v. Dover Area School District* in Pennsylvania. He is also the coauthor of a widely used college textbook, *Biology: Understanding Life*.

Glenn Branch is the deputy director of the National Center for Science Education. He has a BA in philosophy from Brandeis University and an MA in philosophy from UCLA. He has written on evolution education and threats to it for such publications as *BioScience, The American Biology Teacher, Trends in Ecology and Evolution, USA Today,* and *Academe*.

Paul R. Gross is a professor emeritus of life sciences at the University of Virginia, coauthor with Norman Levitt of *Higher Superstition: The Academic Left and Its Quarrels with Science*, and coauthor with Barbara Forrest of *Creationism's Trojan Horse: The Wedge of Intelligent Design*.

Martinez Hewlett is a professor emeritus in the department of molecular and cellular biology at the University of Arizona in Tucson. He has published thirty scientific papers; a novel, *Divine Blood;* and (with Ted Peters) *Evolution: From Creation to New Creation.* He is a founding member of the St. Albert the Great Forum on Theology and the Sciences at the University of Arizona. He serves as an adjunct professor at the Dominican School of Philosophy and Theology at the Graduate Theological Union in Berkeley, California.

The **Reverend Barry W. Lynn** is the executive director of Americans United for Separation of Church and State, an ordained minister in the United Church of Christ, and one of the leaders of the American religious left. He is a strong advocate of separation of church and state and is considered a liberal Christian, both politically and theologically.

Nicholas J. Matzke is public information project director at the National Center for Science Education. Matzke has a double BS in biology and chemistry from Valparaiso University and a master's degree in geography from the University of California, Santa Barbara. He worked extensively with the plaintiffs' legal team in *Kitzmiller v. Dover.*

Ted Peters is an ordained pastor in the Evangelical Lutheran Church in America and professor of systematic theology at Pacific Lutheran Theological Seminary and at the Graduate Theological Union in Berkeley, California. He is author of *God—The World's Future* and *Science, Theology, and Ethics* and coauthor, with Martinez Hewlett, of *Evolution: From Creation to New Creation.* He is editor-in-chief of *Dialog, A Journal of Theology* and coeditor of *Theology and Science,* published by the Center for Theology and the Natural Sciences in Berkeley.

Eugenie C. Scott, a former university professor of physical anthropology, is the executive director of NCSE. She has been both a researcher and an activist in the creationism/evolution controversy for over twenty years. She has received national recognition for her NCSE activities, including awards from the National Science Board, the American Society for Cell Biology, the American Institute of Biological Sciences, the Geological Society of America, and the American Humanist Association. Her book *Evolution vs. Creationism: An Introduction* was named an Outstanding Academic Title for 2005 by *Choice*.

Jay D. Wexler has an MA from the University of Chicago Divinity School and a JD from Stanford Law School. He has been a member of Boston University's School of Law faculty since 2001, specializing in law and religion, on which he has published important articles in a number of law reviews. He has clerked for, among others, Supreme Court Justice Ruth Bader Ginsburg.